Competitive Math
for Middle School

Competitive Math for Middle School

Algebra, Probability, and Number Theory

Vinod Krishnamoorthy

PAN STANFORD PUBLISHING

Published by

Pan Stanford Publishing Pte. Ltd.
Penthouse Level, Suntec Tower 3
8 Temasek Boulevard
Singapore 038988

Email: editorial@panstanford.com
Web: www.panstanford.com

British Library Cataloguing-in-Publication Data
A catalogue record for this book is available from the British Library.

Competitive Math for Middle School: Algebra, Probability, and Number Theory

Copyright © 2018 Pan Stanford Publishing Pte. Ltd.

Cover image: Courtesy of Nirmala Moorthy

For photocopying of material in this volume, please pay a copying fee through the Copyright Clearance Center, Inc., 222 Rosewood Drive, Danvers, MA 01923, USA. In this case permission to photocopy is not required from the publisher.

ISBN 978-981-4774-13-0 (Paperback)
ISBN 978-1-315-19663-3 (eBook)

Contents

Preface

I originally began writing this textbook after teaching creative math to middle school students, who were endlessly fascinated, just as I had been, with the field of mathematics. This book is a compilation of important concepts used in competition mathematics in Algebra, Counting/Probability, and Number Theory. Over 420 problems are provided with detailed solutions found at the end of each chapter. These solutions are intended to guide students to identify a promising approach and to execute the necessary math. I recommend that students try all of the problems, even if they seem intimidating, and use the solutions to the problems as part of the learning process; they are as essential to learning as the teachings and examples given in the body of the text. After reading the solution, students should try to reproduce it themselves. My hope is to provide not only mathematical facts and techniques but also examples of how they may be applied, so that the student gains a thorough understanding of the material and confidence in their problem-solving abilities.

Creative math is becoming increasingly important in schools all over the world. The new trend is conspicuous in the remodeling of standardized tests such as the American SAT, the standard entrance exam for U.S. colleges. Advanced middle school students and high school students can use this book to gain an advantage in school and develop critical thinking skills.

Vinod Krishnamoorthy

Chapter 1

Algebra

Part 1: Linear Equations

Equations are the foundation of mathematics. All forms of math rely on the principle of equality.

An equation states that two expressions have the same value. Expressions are what are on either side of the equation. This may seem obvious, but understanding this is essential.

Variables, or the letters we see in equations, are values that we do not know. We must manipulate the equation to find the value of the variable(s). Coefficients are the numbers located directly left of variables. The coefficient of a variable multiplies the variable's value. For example, $5x$ means "five times x."

Linear equations are the building blocks of algebra. An example of a linear equation is $2x + 3 = 6$. To solve this equation, we must obtain the variable alone on one side and a simplified value on the other. This process is called isolating the variable. It uses principles of inverse operations: subtraction cancels addition, division cancels multiplication, etc.

Example 1: $2x + 3 = 6$. Solve for x.

- To isolate the variable x, we must eliminate the $+3$ and the coefficient 2.

Competitive Math for Middle School: Algebra, Probability, and Number Theory
Vinod Krishnamoorthy
Copyright © 2018 Pan Stanford Publishing Pte. Ltd.
ISBN 978-981-4774-13-0 (Paperback), 978-1-315-19663-3 (eBook)
www.panstanford.com

- To eliminate $+3$, we subtract 3 from the left side of the equation. This leaves us with $2x + 0$, which is the same as just $2x$.

- However, *whatever is done to one side of an equation must also be done to the other*. This is the main rule of solving equations. If it is not followed, the two sides of the equation will no longer be equal.

- Following this rule, we subtract 3 from the right side of the equation as well. We now have $2x = 6 - 3$, or $2x = 3$.

- The next step is to eliminate the 2. Recall that $2x$ means $2 \times x$. Reversing multiplication calls for division, so we divide $2x$ by 2. Doing so leaves us with $1x$, which is equivalent to just x.

- We have to divide the other side of the equation by 2 as well. This leaves us with $x = \dfrac{3}{2}$. The variable is isolated and the other side is simplified, so we are done.

Example 2: $\dfrac{3a + 2}{2} = 6a + 3$. How do we isolate the variable a? When manipulating one side of an equation, always act on the entire side, not just individual parts.

Multiplying both sides of the equation by 2 yields $2\left(\dfrac{3a + 2}{2}\right) = 2(6a + 3)$, or $3a + 2 = 12a + 6$. We were able to simplify the left side in this manner because multiplying $\left(\dfrac{3a + 2}{2}\right)$ by 2 cancels its denominator.

From here we subtract $3a$ from both sides to obtain all terms containing a on one side of the equation. Doing so yields $2 = 9a + 6$. Next, we subtract 6 from both sides, obtaining $9a = -4$. Finally, we divide both sides by 9, obtaining the solution, $a = -\dfrac{4}{9}$.

We can do almost whatever we want to one side of an equation as long as we do the same to the other side. This is because the two sides of an equation by definition hold the same value, and doing the same thing to the same value will always maintain equality.

Example 3: $xy + xy = 4$. Solve for xy.

- Our goal is to isolate the term xy; that is, to obtain xy alone on one side and a simplified value on the other.
- $xy + xy = 2xy$, so the equation can be simplified to $2xy = 4$.
- Dividing both sides by 2, we find that $xy = 2$.

Example 4: If $2(x + 3) = 4$, find the value of $x + 3$.

- To do this, we divide both sides by 2.
- The 2's cancel in $\dfrac{2(x + 3)}{2}$, leaving just $x + 3$. Dividing the right side of the equation by 2 yields $x + 3 = 2$. This is the final answer, as the problem does not ask us to solve for x.

It is important for us to learn to convert words to equations. This technique is essential for solving word problems.

Example 5: Jado has two more than four times as many marbles as Rolf. If Jado has 14 marbles, how many marbles does Rolf have?

- The first step in converting word problems to equations is creating the necessary variables. Let us define the variable r as what the problem asks us to find: the number of marbles that Rolf has.
- If Jado has two more than four times as many marbles as Rolf, Jado has $2 + 4r$ marbles. We also know that Jado has 14 marbles, so $2 + 4r$ is the same as 14. It follows that $2 + 4r = 14$.
- This is now a solvable equation for the variable r.
- First, we subtract two from both sides, obtaining $4r = 12$.
- To eliminate the coefficient 4, we divide both sides by 4. Doing so yields $1r = 3$, which can be rewritten as $r = 3$. This means that Rolf has three marbles.

Note: If given a term such as $\frac{2}{5}a$, the way to remove the coefficient $\frac{2}{5}$ is to multiply by $\frac{2}{5}$'s reciprocal, $\frac{5}{2}$. This will make the term equivalent to $1a$, which is just a.

Example 6: Solve the equation $\frac{7}{5}y - 10 = y$.

- Subtracting y from both sides gives $\frac{2}{5}y - 10 = 0$, and adding 10 to both sides gives $\frac{2}{5}y = 10$.

- How do we isolate y from here? The reciprocal of $\frac{2}{5}$ is $\frac{5}{2}$, and multiplying both sides by $\frac{5}{2}$ yields $\frac{5}{2} * \frac{2}{5}y = \frac{5}{2} * 10$.

- The left side of the equation, $\frac{5}{2} * \frac{2}{5}y$, is equivalent to $\frac{5}{2} * \frac{2}{5} * y$, and from here it is easy to see that this simplifies to $1y$ or just y

- The right side of the equation simplifies to 25, so the final answer is $y = 25$.

Not all equations can be solved.

Let us try to solve $3x + 8 = 4x + 8 - x$.

- Simplifying yields $3x + 8 = 3x + 8$.

- Subtracting 8 from both sides, we obtain $3x = 3x$, and dividing both sides by 3 leaves us with $x = x$.

- Let us think about this. No matter what value x takes on, the equation $x = x$ will hold true. Dividing both sides by x to further simplify the equation, we obtain $1 = 1$, which makes even less sense.

What happened? Regardless of the value of x, this equation is satisfied. Try plugging different values of x into the original equation to see for yourself.

If we had gone from $3x = 3x$ to $1=1$ by dividing both sides by $3x$, we would have a universal truth-a mathematical statement that is always true.

Here is another example: $3x + 8 = 3x + 5$.

Subtracting $3x$ from both sides yields $8 = 5$. No value of x satisfies this equation. It has no solution.

After isolating and simplifying, the general rule is if you end up with a universal truth such as $1 = 1$, $9 = 9$, or $x = x$, all real numbers satisfy the equation, but if you end up with a universal untruth such as $0 = 3$ or $7 = 12$, the equation has no solution.

Problems: Linear Equations

1 Bronze. $x + 4 = 9$. Find x.

2 Bronze. $3x + 3 = 33$. Find x.

3 Bronze. $a + 5 = 8$. Find a.

4 Bronze. $4x + 45 = 49$. Find x.

5 Bronze. $\dfrac{9}{19}x = 18$. Find x.

6 Bronze. $2.5x + 12.1 = 4.6$. Find x.

7 Bronze. $4 + x = 3 + 2x$. Find x.

8 Bronze (calculator). $14.92 + 15.38x = 276.38$. Find x.

9 Bronze. $\dfrac{13}{2}c + 4 = 9 + 10c$. Find c. Express your answer as a common fraction (an improper fraction in lowest terms).

10 Bronze. Joharu and Bebi have 24 coins in total. If Joharu has 18 coins, how many coins does Bebi have?

11 Bronze. Tickets to an amusement park are 5 dollars each. To make 500 dollars in a day, how many people must visit?

12 Silver. Shekar has 22 trading cards and Ashok has 4 less than one third of their combined amount. How many trading cards does Ashok have?

13 Bronze. $(4y + 3) - (2y + 1) = 42$. Find the value of y.

14 Bronze. $(a + 3)/5 = 9$. Find the value of a.

15 Bronze. Flying in an airplane operated by VK airlines costs an initial fee of $100 plus $30 per 150 miles traveled. How far can one fly with $520 to spend?

16 Bronze. Vinod is going to buy a certain number of sheets of paper. He is going to cut each sheet of paper into two half-sheets, and then cut each half sheet into 3 smaller pieces. He needs 84 of the smaller pieces. How many sheets of paper should he buy?

17 Bronze. When a number is doubled and then added to 5, the result is equivalent to one third of the original number. Find the original number.

18 Bronze. $x + 2z = 3 + 2z$. What is the value of x?

19 Bronze. When 12 is subtracted from a number, the result is equivalent to twice the original number. Find the original number.

20 Silver. Solve $x + 12 = 2 \left(\dfrac{x}{2} + \dfrac{3}{2} \right)$.

21 Silver. Solve $2(9 - x) = 4(4.5 - 0.5x)$.

22 Gold. One woman was born on January 1, 1940. Another woman was born on January 1, 1957. They met many years later. When they met, the older woman was one more than twice as many years old as the younger woman. In what year did the two women meet?

23 Bronze. $\dfrac{2x - 4}{7} = 4$. Solve for x.

24 Bronze. $x + 3 = 5$. Find the value of $2x + 9$.

Part 2: Cross Multiplication

Cross multiplication is an important technique for solving equations that contain fractions. In this technique, we multiply both sides of an equation by the denominators of the fractions within it. Doing so will remove the denominators and make the equation easier to solve.

Cross multiplication is based on the principle that for any nonzero values a and b, $a \times \dfrac{b}{a} = b$. In other words, multiplying a fraction by its denominator leaves just its numerator.

Example 1: $\dfrac{2}{x} = \dfrac{1}{5}$.

• First, we multiply both sides by 5, obtaining $\dfrac{10}{x} = 1$.

• Then, we multiply both sides by x. This cancels the denominator of the left side, leaving $x = 10$.

Example 2: $\dfrac{x}{3x+5} = \dfrac{3}{5}$.

- First, we multiply both sides of the equation by 5, obtaining
$$\dfrac{5x}{3x+5} = 3.$$

- Next, we multiply both sides of the equation by $(3x + 5)$. We keep it in parentheses because we are multiplying by the entire expression, not just parts of it.

- This cancels the denominator on the left side, leaving us with $5x = 3(3x + 5)$ or $5x = 9x + 15$.

- Solving, we find that $x = -\dfrac{15}{4}$.

The following example does not involve cross multiplication, but it involves fractional expressions with variables.

Example 3: Simplify $\dfrac{a}{1 + \dfrac{4}{5}a}$ so that there are no fractions within the larger fraction.

- Recall that for all nonzero values of x and y, $\dfrac{x}{y} = x \times \dfrac{1}{y}$. However, $\dfrac{x}{y+z}$ cannot be simplified further.

- To apply the first principle on the given expression, we must collapse the denominator into a single term. This can be done using common denominators.

- $1 = \dfrac{5}{5}$, and $\dfrac{4}{5}a = \dfrac{4a}{5}$. Therefore, $1 + \dfrac{4}{5}a = \dfrac{5+4a}{5}$.

- $\dfrac{a}{\dfrac{5+4a}{5}} = a \times \dfrac{5}{5+4a} = \dfrac{5a}{4a+5}$.

Example 4: Simplify $\dfrac{a}{\dfrac{3}{2} + \dfrac{4}{3}a}$ so that there are no fractions within the larger fraction.

- Since the fact that $\dfrac{x}{y} = x \times \dfrac{1}{y}$ can only be applied if the denominator is a single term, we must find a common denominator for the two terms in the sum.

- $\dfrac{3}{2} = \dfrac{9}{6}$ and $\dfrac{4}{3}a = \dfrac{8a}{6}$. Therefore, $\dfrac{3}{2} + \dfrac{4}{3}a = \dfrac{9 + 8a}{6}$.

- $\dfrac{a}{\frac{9 + 8a}{6}} = a \times \left(\dfrac{6}{9 + 8a}\right) = \dfrac{6a}{8a + 9}$.

Note: Dividing any value (including 0) by 0 is mathematically undefined. When doing arithmetic and solving equations, dividing by expressions equivalent to 0 can often lead to incorrect results. Always be wary of this.

Problems: Cross Multiplication

1 Bronze. $\dfrac{3}{2x} = \dfrac{2}{8}$. Find x.

2 Bronze. $\dfrac{4}{x} = 10$. Find x. Express your answer as a common fraction.

3 Silver. $\dfrac{9}{9 + x} = \dfrac{12}{19 + x}$. Find x.

4 Silver. $\dfrac{4(3 + x)}{3(9 - x)} = \dfrac{8}{3}$. Find x.

5 Silver. A worker put 200 gallons of water into one tank and a certain amount of water into a second tank. The combined amount of water from the two tanks flowed to a filter, which removed half of the water. The remaining water went to a production factory, but here its amount was tripled. There are now 939 gallon of water in total. How much water did the worker originally put into the second tank?

6 Bronze. $\dfrac{1}{2z} + \dfrac{5}{6z} = 12$. Find z.

7 Silver. Simplify $\dfrac{2x}{\frac{x}{3} + \frac{y}{2} + \frac{1}{5}}$.

Part 3: Systems of Equations

As you may have already realized, a single equation with two different variables cannot be solved. Two different variables cannot be combined with addition or subtraction. For example, the expression $x + y$ cannot be simplified any further. But with two equations using the same two variables, it is possible to find the values of both.

A set of multiple equations with the same variables is called a system of equations. In most cases, for a system to be solvable, the number of variables must be less than or equal to the number of equations given. An example of a system of equations is

$$2x + y = 3$$

$$3x + y = 5$$

The first method used to solve systems of equations is substitution. In this method, we use one equation to find a variable *in terms of* the other(s), and then substitute what the variable is equivalent to into the other equation(s).

Here is a quick example to present the concept of substitution.

Example 1: $y = 34$ and $x + y = 21$. Find x.

- Since $y = 34$, the variable y has a value of 34. y and 34 are interchangeable in any equation.

- By this logic, we can replace y with 34 in $x + y = 21$ to form the equation $x + 34 = 21$.

- Solving this equation, we find that $x = -13$.

To solve for b *in terms of* a in the equation $3b + a = 12$, we must isolate the variable b. We want the right side of the solution to be an expression containing a.

Subtracting a from both sides yields $3b = 12 - a$, and dividing both sides by 3 yields $b = 4 - \frac{1}{3}a$. We successfully solved for b in terms of a, as we have an expression containing a that is equivalent to b.

Example 2: Let us solve the previous example: $2x + y = 3$ and $3x + y = 5$. We will go through solving for y in terms of x, but either way works fine.

- We start with the first equation, $2x + y = 3$. To solve for y in terms of x, we simply isolate y and ignore the variable on the other side.

- We isolate y by subtracting $2x$ from both sides. This leaves us with $y = 3 - 2x$. We just solved for y in terms of x.
- Next, we substitute the expression on the right side into the second equation: $3x + y = 5$. Since y is equivalent to $3 - 2x$, we can replace y with $3 - 2x$. Doing so yields $3x + (3 - 2x) = 5$.
- This is a solvable equation for the variable x. Solving, we find that $x = 2$.
- We are not done yet, as we still have to find the value of y. Since we already know the value of x, we can replace x with 2 in $2x + y = 3$. In doing this we reuse one of the original equations given to us in the problem.
- $2(2) + y = 3$ is a solvable equation for y where $y = -1$.

Solving systems of equations is not the only application of substitution. Substitution can be used in many scenarios where you have multiple pieces of information.

The second method used to solve systems of equations is called elimination. Elimination is the process of adding the equations to take away a variable and receive a simpler equation in return. To do this, one or more of the equations must be manipulated such that their sum cancels a variable.

Example 3: $2x + y = 3$ and $3x + y = 5$.

- First, we choose a variable to eliminate. As you will soon see, choosing the variable y will be easier.
- We now set up the variable to be eliminated. To do this, we multiply both sides of the second equation by -1. This yields $-3x - y = -5$. Since we did the same thing to both sides, the equation still holds true.
- Next, we add the two equations: $(2x + y) + (-3x - y) = 3 - 5$. Notice how y and $-y$ cancel to 0: this is the goal of elimination.
- Simplifying yields $-x = -2$, so $x = 2$. Plugging the value of x back into the first equation as we did in the previous example, we find that $y = -1$.

Example 4: $5x + 5y = 12$ and $3x + 2y = 7$.

- Let us use elimination to solve this system. How do we set up one of the variables to be eliminated?
- One way is to make the coefficient of y 10 in the first equation and -10 in the second. 10 is divisible by both 2 and 5, so both equations can be multiplied by integers to make the coefficients of y 10 and -10. This ties into the concept of *least common multiples*, which is explained in chapter 3.
- Know that the reason we chose 10 and -10 was to make the following steps easier, but any number and its negative can be used in elimination.
- In the first equation, we multiply both sides by 2. This leaves $10x + 10y = 24$.
- To make the coefficient of y -10 in the second equation, we multiply both sides by -5. This leaves $-15x - 10y = -35$.
- Now, we add the two equations: $(10x + 10y) + (-15x - 10y) = 24 - 35$. Simplifying yields $-5x = -11$. It follows that $x = \dfrac{11}{5}$.
- Plugging the value of x back into $3x + 2y = 7$ yields $\dfrac{33}{5} + 2y = 7$. Therefore, $2y = \dfrac{2}{5}$ and $y = \dfrac{1}{5}$.

Notice how we added the equations in the elimination examples. Why are we able to do this?

- Consider the two equations $a + b = c$ and $d - e = f$. We want to show that the equation $(a + b) + (d - e) = c + f$ is valid.
- Let us start with $a + b = c$. To obtain $(a + b) + (d - e)$, we add $(d - e)$ to both sides. This leaves us with $(a + b) + (d - e) = c + (d - e)$.
- However, we know that $d - e = f$. Therefore, we can replace $(d - e)$ with f on the right side to form the desired result, $(a + b) + (d - e) = c + f$.

What we showed is one example of the property of adding equations, but the principle can be used to create a generic proof for all equations.

No one method is better than the other. As to which method is easier depends on the system being solved and your personal preference. Regardless of the method chosen, the answer will be the same.

Some problems ask for the answer as an *ordered pair*. Ordered pairs are pairs of values put in the form (_, _).

Consider a problem that asks for the answer as an ordered pair (a, b). Instead of writing "$a = 1$ and $b = -2$," one would write $(1, -2)$.

In ordered pairs, order matters. For example, $(1, 2)$ is completely different than $(2, 1)$. If the variables in a problem are x and y, the assumed form of the ordered pair is (x, y), and if the variables in a problem are x, y, and z, the assumed form of the ordered *triple* is (x, y, z).

Problems: Systems of Equations

1 Silver. $x + y = 7$ and $2x + y = 11$. Find x and y.

2 Bronze. $x + y = 50$ and $x - y = 10$. Find x and y.

3 Silver. $2x + y = 20$ and $x + 2y = 16$. Find x and y.

4 Silver. $3a + 5b = 12$ and $\frac{3}{2}a + \frac{1}{2}b = \frac{5}{2}$. Find a and b.

5 Silver. In a store, two pencils and three pens cost 58 cents. Four pencils and four pens cost 100 cents. What is the cost of one pencil?

6 Silver. In a grocery store, 16 bottles of water and 37 bottles of syrup cost 90 dollars. 8 bottles of water and 23 bottles of syrup cost 54 dollars. If both bottles of water and bottles of syrup have constant costs, find the cost of one bottle of water.

7 Gold. $a + b = 6$, $b + c = 9$, and $a + c = 11$. Find c.

8 Bronze. $xy + xy = 32$. Find the value of xy

9 Silver. Camels in San Diego zoo have either one hump, two humps, or three humps. There are exactly 10 camels with two·

humps, and there are 21 camels and 45 humps in total. How many camels with one hump are in the zoo?

10 Gold. Linear equations can be written in the form $y = mx + b$, where m and b are constants but x and y can vary. An equation of this form can be thought of as an infinitely large set of solvable equations, each with one solution for y depending on the value of x. If y is 10 when x is 4 and y is 19 when x is 7, find the values of m and b.

11 Silver. $\dfrac{30}{y} = \dfrac{25}{x}$ and $x + 2y = 85$. Find x and y.

12 Gold. $ab + cd = ef$. Solve for c in terms of $a, b, d, e,$ and f.

13 Gold. $c(c + d) = 13$. Solve for d in terms of c.

14 Bronze. $abcd + 4abcd = 10$. Find the value of $abcd$.

Part 4: Exponents and Roots

Exponents signify that a term is to be multiplied by itself over and over again. An example of a term with an exponent is 2^4. The base, or the number in regular script, is the number being multiplied by itself. The exponent, or the number in superscript, is the number of times that the base multiplies itself.

Just as $2 \times 4 = 2 + 2 + 2 + 2 = 8$ and $5 \times 2 = 5 + 5 = 10$, $2^4 = 2 \times 2 \times 2 \times 2 = 16$ and $5^2 = 5 \times 5 = 25$. Exponents work with variables too: $x^4 = x \times x \times x \times x$ and $n^m = n \times n \times n \dots$ (m times). x^y is said "x to the power y" or "x raised to the power of y."

The principle of exponents can be used in reverse. $x \times x$ simplifies to x^2, and $3x \times x$ simplifies to $3x^2$.

Exponents work with more than just single terms. For example, $(x + 1)^2$ can be expanded as $(x + 1)(x + 1)$. We will learn later where to go from here.

When the same variables are raised to different exponents, they cannot be added or subtracted. For example, neither $x^4 + x^5$ nor $4x + 2x^2$ can be added in any way.

Exponents have precedence over addition, subtraction, multiplication, and division in the order of operations. Therefore, an expression such as $2x^3$ cannot be simplified further. This is because $2x^3$ does not equal $2x \times 2x \times 2x$; instead, it equals $2 \times (x \times x \times x)$.

Also, an expression such as -2^4 equals $-(2 \times 2 \times 2 \times 2) = -16$, since the negative sign acts as multiplication by -1. However, $(-2)^4 = (-2)(-2)(-2)(-2) = 16$, because operations in parentheses must be done first. For this reason, expressions in parenthesis are acted on as a whole and are not split up.

Example 1: Simplify $(ab)^2 + ab^3$.

- $(ab)^2 = ab \times ab = a \times a \times b \times b = a^2b^2$, since parentheses indicate that the enclosed operation must be done first. Because of the parenthesis, ab is acted on as a unit.

- ab^3 cannot be simplified further. $ab^3 = a \times (b^3)$, since exponents precede multiplication in the order of operations.

- The final answer is $a^2b^2 + ab^3$.

Example 2: Simplify $-(-2)^4 \times x + 3x^2$.

- Due to the order of operations, we take care of the exponents in the expression first. Since -2 is enclosed in parentheses, -2 is raised to the power 4, not just 2. $(-2)^4 = (-2)(-2)(-2)(-2) = 16$.

- The first term simplifies to $-16x$ because of the negative sign in front of $(-2)^4$ and x being multiplied afterwards.

- The expression is now $-16x + 3x^2$, which cannot be simplified further.

There are four important rules to remember when combining exponential terms (a.k.a. terms containing exponents).

(1) The product of two or more exponential terms with the same base equals the common base raised to the sum of the exponents.

$$x^y \times x^z = x^{y+z}.$$

- Consider the product $x^2 \times x^3$. Expanding x^2 and x^3 yields $(x \times x) \times (x \times x \times x)$. The parentheses are irrelevant in this multiplication, so $x^2 \times x^3 = x^{2+3} = x^5$.

(2) The quotient of two or more exponential terms with the same base equals the common base raised to the difference of the exponents.

$$x^y / x^z = x^{y-z}.$$

- Consider x^4 / x^2. Expanding both terms yields $\dfrac{x \times x \times x \times x}{x \times x}$. The two instances of x in the denominator cancel with exactly two instances of x in the numerator, so $\dfrac{x^4}{x^2} = x^{4-2} = x^2$.

(3) An exponential term raised to a certain power is equivalent to the base of the exponential term raised to the product of the exponents.

$$(x^y)^z = x^{y \times z}.$$

- Consider $(x^2)^4$. Expanding yields $(x \times x) \times (x \times x) \times (x \times x) \times (x \times x)$. The parentheses are irrelevant in this multiplication, so $(x^2)^4 = x^{2 \times 4} = x^8$.

(4) When the product of multiple terms is raised to a certain power, expand the expression by raising each term to the specified power and multiplying the results.

$$(wxyz)^a = w^a \times x^a \times y^a \times z^a.$$

- Consider $(abc)^3$. Expand it into $(a \times b \times c) \times (a \times b \times c) \times (a \times b \times c)$. Due to the Commutative Property of Multiplication, we can rewrite this as $a \times a \times a \times b \times b \times b \times c \times c \times c$, or $a^3 b^3 c^3$.

If two exponential terms have different bases, they cannot be combined in any of these ways.

Based on the above rules and the fact that any term or expression divided by itself is equal to 1, we can discover that anything raised to the power of 0—numbers, variables, and even full expressions— is equal to 1.

(*Note:* x is equivalent to x^1. This is important to remember when working with exponents).

Taking a root of a number is the opposite of raising it to a power. To take the n-th root of x, find the number that when raised to the n-th power equals x. For example, the third root of 8 is 2, since $2^3 = 8$. The n-th root of a number is equivalent to the number raised to the power of $\frac{1}{n}$. For example, the 3-rd root of 8 is equivalent to $8^{\frac{1}{3}}$.

When taking a root of a fraction, the root distributes to the numerator and denominator.

Example 3: Find the third root of $\frac{8}{27}$.

- The third root of $\frac{8}{27}$ equals the third root of 8 divided by the third root of 27.

- This is because the product of fractions equals the product of their numerators divided by the product of their denominators. If we call the third root of $\frac{8}{27}$ $\frac{x}{y}$, $\left(\frac{x}{y}\right)^3 = \frac{x}{y} \times \frac{x}{y} \times \frac{x}{y} = \frac{x^3}{y^3}$.

- The third root of 8 is 2 and the third root of 27 is 3, so the answer is $\frac{2}{3}$.

Roots can also be represented by a $\sqrt{}$ symbol called a radical sign. The number inside the radical sign is the number whose root is being taken, and the number in superscript on the left of the symbol is the degree of the root being taken (third root, fourth root, etc.). If there is no number in superscript on the left of the radical sign, take the second root a.k.a. the square root of the inside number.

- $\sqrt{x}$: square root
- $\sqrt[3]{x}$: cube root
- $\sqrt[4]{x}$: fourth root
- Etc.

Rational numbers can be expressed as integers or fractions. Their decimal representations either terminate (end somewhere) or repeat indefinitely in a certain pattern. Most roots turn out to be irrational—their decimal representations continue on forever without any pattern. Only some roots simplify to rational numbers.

There is no straightforward formula to find the rational roots of numbers; common methods include guess and check and memorization.

The roots of integers that exist are either integers as well or irrational numbers—they never simplify to terminating or repeating decimals. This fact is easy to observe but a little more difficult to definitively prove. Simple locating techniques can be used to solve for the roots that are rational and determine which ones are irrational.

Example 4: Determine whether $\sqrt{22}$, $\sqrt{81}$, and $\sqrt{185}$ are irrational or rational.

- 22 is an integer, so $\sqrt{22}$ is either a positive integer or irrational. Note that $\sqrt{22}$ is positive, but there are two square roots of 22: one is positive and the other is negative. Since $4^2 = 16$ and $5^2 = 25$, the positive square root of 22 is between 4 and 5, as 22 is between 16 and 25.

- There are no integers between 4 and 5, so $\sqrt{22}$ is not an integer. Therefore, it is irrational.

- $10^2 = 100$, so $\sqrt{81}$ is less than 10. Testing the squares of positive integers going backwards from 10, we find that $9^2 = 81$, and therefore $\sqrt{81} = 9$ and is rational.

- If $\sqrt{185}$ is rational, it must be equivalent to a positive integer. If $\sqrt{185}$ falls between two positive integers, then it is irrational.

- What positive integers is $\sqrt{185}$ around? 10^2 and 20^2 are both easily recognizable as 100 and 400, respectively, so from this we can deduce that $\sqrt{185}$ is somewhere between 10 and 15. The fact that $15^2 = 225$ confirms this assumption.

- Going backwards from there, $14^2 = 196$ and $13^2 = 169$. Since 185 is between 169 and 196, $\sqrt{185}$ is between 13 and 14. There are no integers between 13 and 14, so $\sqrt{185}$ is irrational.

This locating technique can also be used to approximate square roots. For example, $\sqrt{22}$ can first be determined to be between 4 and 5. Then, extending this method out to one more digit, $\sqrt{22}$ can be determined to be between 4.6 and 4.7. Of course, determining this for yourself will take some work. Going even further, $\sqrt{22}$ can be determined to be between 4.69 and 4.70, though this may take a lot of tedious multiplication. However, we now have a fairly good approximation of the square root of 22.

Every root of even degree of a positive number has two answers: one positive root and its negative correspondent. There are actually two square roots of 4: 2 and -2. This is because both 2^2 and $(-2)^2$ equal 4.

However, radical signs imply solely the positive solution. We would denote the positive square root of x as $\sqrt{x}$ and the negative square root of x as $-\sqrt{x}$. Similarly, we would denote the positive fourth root of x as $\sqrt[4]{x}$ and the negative fourth root of x as $-\sqrt[4]{x}$.

Roots of odd degree do not follow the above rule: they only have one solution. For example, -2 is not a cube root of 8 because the cube of -2 is -8.

Example 5: Find $27^{\frac{2}{3}}$.

- We will split the exponent into two different exponents that we know how to work with, and then solve from there.

- $27^{\frac{2}{3}}$ can be split into $(27^{\frac{1}{3}})^2$. $27^{\frac{1}{3}} = 3$, so the final answer is 3^2 or 9.

- We can also use the fact that $x^y \times x^z = x^{y+z}$ to split $27^{\frac{2}{3}}$ into $27^{\frac{1}{3}} \times 27^{\frac{1}{3}}$, which simplifies to 9 as well.

Higher order exponents problems can be simplified in numerous ways.

Example 6: Find the sixth root(s) of 729.

- If we call the solution x, $x^6 = 729$. The way to directly isolate x is to take the sixth root of both sides.

- Finding sixth roots is significantly more difficult than finding square roots. Let us first raise both sides to the $\dfrac{1}{2}$ power. Doing so yields $(x^6)^{\frac{1}{2}} = 729^{\frac{1}{2}}$, which simplifies to $x^3 = 27$ and $x^3 = -27$.

- These equations are easy to solve. We find that the sixth roots of 729 are 3 and -3.

(The square roots of negatives are not real numbers. For now, regard taking the square root of a negative number as impossible.)

Example 7: $(x^4)^2 = 256$. Find x.

- As we learned before, $(x^4)^2$ is equivalent to x^8. Now we have $x^8 = 256$.
- Raising both sides to the power 1/2 yields $x^4 = \pm 16$. Since -16 has no square root, $x^4 = 16$ is all that remains. Raising both sides to the power 1/2 again yields $x^2 = \pm 4$.
- Repeating this process one last time yields $x = \pm 2$. The $\pm$ sign is used to denote that there are two values: the positive value and its negative correspondent.

Consider the term x^{-1}. Since $\dfrac{x^y}{x^z} = x^{y-z}$, $x^{-1} = \dfrac{x^0}{x^1}$ or $\dfrac{1}{x}$. Since $(x^y)^z = x^{y \times z}$, $x^{-n} = (x^{-1})^n = \left(\dfrac{1}{x}\right)^n = \dfrac{1}{x^n}$ for all real n.

It follows that a number raised to a negative exponent is equivalent to the reciprocal of the number raised to the positive correspondent of the exponent. $x^{-n} = \left(\dfrac{1}{x}\right)^n = \dfrac{1}{x^n}$.

For example, $2^{-4} = \left(\dfrac{1}{2}\right)^4 = \dfrac{1}{16}$.

Example 8: Simplify $\dfrac{x^4 y^6 z^3}{x^2 y^4 z^6}$.

- The first step is splitting the expression into $\left(\dfrac{x^4}{x^2}\right)\left(\dfrac{y^6}{y^4}\right)$ $\left(\dfrac{z^3}{z^6}\right)$. Make sure you know why doing this is valid.

- Next, we simplify all three quotients. $\dfrac{x^4}{x^2} = x^{4-2} = x^2$, $\dfrac{y^6}{y^4} = y^{6-4} = y^2$, and $\dfrac{z^3}{z^6} = z^{3-6} = z^{-3} = 1/z^3$.

- The product of these is $\dfrac{x^2 y^2}{z^3}$.

Problems: Roots and Exponents

1 Bronze. What is the value of 9^2?

2 Silver. What is the value of $-3^4 + (3/2^2)$? Express your answer as a common fraction.

3 Bronze. If $x^2 = 64$, what are the two possible values of x?

4 Bronze. $140^{23} \times 140^x = 140^{35}$. Find x.

5 Bronze. Find the cube root of 8. (cube root means third root)

6 Bronze. $a^5 = -1$. Find a.

7 Bronze. $a^5 = -32$. Find a.

8 Bronze. Simplify $(3xy)^2$.

9 Bronze. Simplify $\sqrt{144}$.

10 Bronze. $a^4 = 81$. Find the positive value of a.

11 Bronze. $x^3 = 4x$. Find the positive value of x.

12 Bronze. $7x^{99} = x^{100}$. Find the positive value of x.

13 Silver. $x^2 = 2585214$. Find the sum of the possible values of x.

14 Bronze. Find the positive value of x if $x^2 = 169$.

15 Silver (calculator). Find $\sqrt[4]{1296}$.

16 Bronze. Simplify $\dfrac{x^2 y^2}{x^3 y^5}$.

17 Bronze. Simplify $\dfrac{a^3 b^2 c^4}{a^{-3} b^4 c^{-2}}$.

18 Bronze. Simplify $\dfrac{3 \times 9^8}{2 \times 9^5}$.

19 Silver. How many positive integers are greater than the square root of 50 but less than the square root of 150?

20 Silver. Vinod has two sheets of paper. He cuts each sheet into three pieces. He then cuts each piece into 3 smaller pieces. He repeats the cutting process 4 more times. How many pieces of paper does he end up with?

21 Bronze. Find all possible values of $9^{-1/2}$.

22 Bronze. Find all possible values of $16^{3/4} \times 16^{1/2}$.

23 Bronze. $n@m$ means $n^2/(m^2 + n)$. Find 4@5.

24 Bronze. Find the value of $-\left(\dfrac{6}{2}\right)^4$.

25 Bronze. Is $(-2)^{155}$ positive or negative?

26 Bronze. Simplify $(\sqrt[3]{25})(\sqrt[3]{5})$.

27 Bronze. Simplify $\sqrt{\dfrac{625}{9}}$.

28 Bronze. $2n^2 + 8 = 16$. Solve for n.

29 Silver. $\dfrac{2}{3}a^{25} = 162$. Find the value of a^5.

30 Silver. $64^{2x} = 8^{11}$. Find x.

31 Silver. $\dfrac{(x+3)^2 + 19}{2} = 25$. Solve for x.

32 Silver. $\dfrac{(x-3)^5 + 3}{5} = 7$. Solve for x.

Part 5: Simplifying Radical Expressions

Perfect squares are positive integers that are the squares of other positive integers. 1, 4, 9, 16, 25, 36... are the perfect squares. The same goes for perfect cubes, which are whole numbers that are the cubes of other whole numbers, and so forth with perfect fourth powers, perfect fifth powers, etc.

Just like fractions, radicals have to be written in their simplest form the majority of the time.

We have previously learned that $\sqrt{xy}$ can be split into $\sqrt{x} \times \sqrt{y}$. Therefore, an expression like $\sqrt{x^3}$ can be split into $\sqrt{x^2} \times \sqrt{x}$. $\sqrt{x^2}$ equals x, so $\sqrt{x^3}$ can be written as $x\sqrt{x}$. (Note that all of this works only if the values are positive because square roots of negatives are not real numbers).

Similarly, an expression such as $\sqrt{x^2 y^5}$ can be split into $\sqrt{x^2} \times \sqrt{y^5}$. $\sqrt{x^2} = x$, and $\sqrt{y^5} = \sqrt{y^4}\sqrt{y} = y^2\sqrt{y}$.

This is how radicals are simplified; all perfect powers that correspond to the degree of the root are taken out and written as coefficients.

Example 1: Simplify $\sqrt{80}$.

- What factors of 80 are perfect squares? 16 is a factor of 80, and $\sqrt{16} = 4$, so $\sqrt{80}$ can be split into $\sqrt{16} \times \sqrt{5} = 4\sqrt{5}$.

- 5 has no perfect square factors except for 1—and if we factor out 1 nothing will happen—so this is in simplest form.

Example 2: Simplify $\sqrt{92}$.

- One perfect square factor of 92 is 4. $\sqrt{92} = \sqrt{4} \times \sqrt{23} = 2\sqrt{23}$. 23 has no perfect square factors greater than 1, so we are done simplifying.

- Alternatively, instead of only testing from the pool of known factors of 92, we could try dividing 92 by known perfect squares (4, 9, 25, ...) to test whether the radical can be simplified by taking out one or more of them. This is just another way of coming up with perfect square factors.

Note: You don't have to get to the answer right away. For example, if you don't notice that in $\sqrt{72}$ the 72 is divisible by 36, pulling out a 4 and then a 9 works just as well.

$$\sqrt{72} = \sqrt{4} \times \sqrt{18} = 2\sqrt{18} \quad = 2\left(\sqrt{9} \times \sqrt{2}\right) = 2 \times 3\sqrt{2} = 6\sqrt{2}$$

Problems: Simplifying Radical Expressions

1 Silver. Simplify $\sqrt{x^4 y^3}$.

2 Bronze. Write $\sqrt{84}$ in simplest form.

3 Silver. Write $\sqrt{288}$ in simplest form.

4 Silver. Write $\sqrt{182}$ in simplest form.

5 Gold. Write $\sqrt[3]{54}$ in simplest form.

6 Gold. Write $\sqrt[4]{810}$ in simplest form.

7 Silver. Find the perfect square closest to 950.

Part 6: Proportions

Ratios, also known as proportions, are fractions often used to express the relationships between different things. The ratio of apples to bananas means the value of the number of apples divided by the number of bananas.

Let us say that we have 6 apples and 10 bananas. The ratio of apples to bananas is $\frac{6}{10}$ or $\frac{3}{5}$.

Ratios can be denoted in many ways. The ratio $\frac{3}{5}$ can also be written 3:5 or "3 to 5."

Let us say that the ratio of boys to girls in a classroom is 1:2. This also means that there is *one boy for every two girls* in the classroom. There are many different numbers of boys and girls that satisfy this, but the ratio always simplifies to 1:2. Some combinations that work are 100 boys and 200 girls, 1 boy and 2 girls, and 43 boys and 86 girls.

If given the ratio of two different things and the actual amount of one, it is possible to solve for the amount of the other.

Example 1: The ratio of men to women at a concert is 3 to 5. If there are 330 men, how many women are at the concert?

- Since the ratio of men to women is $\frac{3}{5}$, we write the equation $\frac{m}{w} = \frac{3}{5}$. m is the total number of men and w is the total number of women. There are 330 men at the concert, so $m = 330$.

- From these two equations, we find that $w = 550$, so there are 550 women at the concert.

Proportions are hidden in many problems.

Example 2: If one bird lays two eggs in a year, how many eggs do two birds lay in a year?

- Since 1 bird lays 2 eggs, the ratio of the number of birds to the number of eggs is $\frac{1}{2}$. This ratio will stay constant even if the number of birds or the number of eggs changes, because for every one bird there should be two eggs laid per year.

- Now that we defined the ratio, we set up the equation $\frac{\text{birds}}{\text{eggs}} = \frac{1}{2} = \frac{2}{x}$.

- Solving this equation, we find that $x = 4$, so two birds lay four eggs in a year.

The two previous problems are examples of direct proportionality. Direct proportionality between variables happens when as one increases, the other increases as well. However, the ratio or quotient of the two variables always stays constant.

Example 3: x is directly proportional to y, and $x = 3$ when $y = 6$. What is the value of x when $y = 20$?

- First, we must find the value of $\frac{x}{y}$. This is $\frac{3}{6}$ or $\frac{1}{2}$. This will remain true no matter what the value of y is.

- We now have two pieces of information: $\frac{x}{y} = \frac{1}{2}$ and $y = 20$.

- Solving this system of equations for x, we find that $x = 10$.

Inverse proportionality is the opposite of direct proportionality. Instead of the quotient, the product of the variables is constant. This means that as one variable decreases, the other increases.

Example 4: x and y are inversely proportional. If x is 3 when y is 6, find x when y is 9.

- $x \times y$ has a constant value. This value equals $3 \times 6 = 18$.

- When y is 9, $x \times y$ still equals 18. Therefore, x equals 2.

More than two variables can be related in this manner. If a is directly proportional to b and c, the quotient of the three variables

$(a/b/c)$ is constant. $a/b/c$ can be written as $\dfrac{a}{b} \div c$, which equals $\dfrac{a}{b} \times \dfrac{1}{c}$ or $\dfrac{a}{bc}$. We can see that b and c are inversely proportional within the larger proportion, as the term bc is in the fraction. Therefore, bc has a constant value as long as a remains constant.

If a is inversely proportional to b and directly proportional to c^2, ab/c^2 has a constant value.

Example 5: If $\dfrac{ab}{cd}$ has a constant value, what happens to b as c increases and a and d stay constant?

- $\dfrac{b}{c}$ is nested within the larger proportion. Therefore, b is directly proportional to c. As c increases, b increases as well, as long as the other two variables are not doing anything to change that.

- Prove that this is true: Let us call the constant value of $\dfrac{ab}{cd}$ x. If $\dfrac{ab}{cd} = x$, $\dfrac{b}{c} = \dfrac{x \times d}{a}$. If a and d remain constant, $\dfrac{x \times d}{a}$ will also remain constant, since x is a constant as well. Therefore, $\dfrac{b}{c}$ has a constant value, and b is directly proportional to c.

Example 6: If 10 faucets fill 20 tubs in 2 days, how many tubs will 15 faucets fill in 5 days?

- With our current knowledge of proportions, it is usually easy to tell whether proportions can be used in a problem.

- Since it mainly deals with rates, this problem definitely calls for the use of proportions. Whether it involves inverse or direct proportionality is the question. Let us call the number of faucets f, the number of tubs they can fill t, and the number of days they work for d.

- As the number of faucets increases, the number of tubs they can fill increases as well, so f is directly proportional to t. Also, as the number of faucets increases, the number of days they work for decreases, so f is inversely proportional to d.

- Since f is directly proportional to t and inversely proportional to d, $\dfrac{fd}{t}$ has a constant value.
- We are given that when f is 10, t is 20 and d is 2. It follows that the constant value $\dfrac{fd}{t} = \dfrac{10(2)}{20}$ or 1.
- When $f = 15$ and $d = 5$, $\dfrac{fd}{t} = \dfrac{15(5)}{t}$ or $\dfrac{75}{t}$. $\dfrac{fd}{t}$ always equals 1, so $\dfrac{75}{t} = 1$. Solving this equation, we find that $t = 75$ tubs.

Let us say that a box has two types of things, and the ratio of the amount of the first thing to the amount of the second thing is $a:b$. The fraction of the *total box* that is the first thing is $\dfrac{a}{a+b}$, not $\dfrac{a}{b}$, since the total box includes both things, not just the second.

Example 7: If a mixture is made up of 4 liters of water and 10 liters of oil, the ratio of water to oil in the mixture by volume is $\dfrac{4}{10}$ or $\dfrac{2}{5}$, but the fraction of the total mixture that is water is $\dfrac{4}{10+4}$ or $\dfrac{2}{7}$.

Always remember that proportions are fractions, so proportions follow all rules that fractions do.

Problems: Proportions

1   Bronze. x is inversely proportional to y. If x is multiplied by 3, what does y become in terms of y's original value?

2   Bronze. If 5 birds lay 10 eggs every month, how many eggs do 10 birds lay in a month?

3   Bronze. If 16 birds lay 24 eggs in a month, how many eggs do 20 birds lay in a month?

4   Silver. x is inversely proportional to y and directly proportional to z. If $x = 1$ when $y = 4$ and $z = 5$, what is x when $y = 10$ and $z = 12$?

5   Gold. If 3 workers can do 2 jobs in 5 days, how long will it take for 4 workers to do 6 jobs?

6 Silver. The square of the length of an alien's hair is inversely proportional to its height. If a 9-inch tall alien has 9-inch long hair, what is the length of the hair of a 4-inch tall alien? Express your answer as a decimal to the nearest tenth.

7 Silver. The ratio of x to y is 2:5. If $2x + 3y = 57$, find y.

8 Bronze. a and b are directly proportional. If a is 3 when b is 1, what is a when b is 98?

9 Silver. a is directly proportional to b and inversely proportion to c^2. If a is 21 when b is 12 and c is 2, then what is a when $b = 28$ and $c = 7$?

10 (calculator) Bronze. A human weighing 100 pounds on Earth will weigh 38 pounds on Mars. How much will a human weighing 100 pounds on Mars weigh on Earth? Express your answer to the nearest pound.

11 Silver. The number of plants in a house is directly proportional to the cube of the amount of oxygen in it. If a house at one point has 8 plants and 2 pounds of oxygen, how many pounds of oxygen will be in the house when it has 27 plants?

12 Silver. A box contains 90 pencils, each of which is either red or blue. The ratio of red pencils to blue pencils is 7:11. How many blue pencils are in the box?

13 Gold. Two bags each contain some number of pencils and erasers. In the first bag, the ratio of pencils to erasers is 9:2. In the second bag, the ratio of pencils to erasers is 6:5. The contents of both bags are then dumped into an empty box. If the box contains 57 pencils and 31 erasers, how many total items were originally present in the second bag?

14 Silver. In the planet Vinoe, the ratio of land to water in square miles is 3:5. In Vinoe's moon Eoniv, the ratio of land to water in square miles is 2:9. If there are 752 square miles in total on Vinoe and 374 on Eoniv, then what is the combined ratio of water to land between both the planet and the moon?

15 Silver. The dimensions of a community on its blueprint are proportional to those of the actual community. If a 45 foot-long

street is 9 inches on the blueprint, how long is a street that is 12 inches on the blueprint?

16 Silver. The ratio of c to d is 9:23. What is $4c:5d$?

Part 7: Inequalities

Equations are useful when a variable has only one possible value. But what if a single variable has many possible values? This is where inequalities come into play. Inequalities are just like equations, except instead of the $=$ sign, they use $<$, $>$, $\leq$, and $\geq$.

The $<$ symbol translates to "less than." If $x < 3$, x is less than 3.

The $>$ symbol means "greater than." If $x > 3$, x is greater than 3.

The symbol $\leq$ means "less than or equal to." If $x < 3$, x cannot equal 3. However, if $x \leq 3$, x can equal 3.

Lastly, $\geq$ means "greater than or equal to," with properties similar to the previous symbol.

Inequalities can be read either way. For example, $3 < x$ can also be read $x > 3$ if it is read from right to left instead of left to right. Both mean exactly the same thing.

Inequalities work similarly to equations. If $x + 3 < 5$, you can subtract 3 from both sides, obtaining $x < 2$. If $2x + 3 \geq 7$, you can subtract 3 from both sides, obtaining $2x \geq 4$, and then you can divide both sides by 2 to find that $x \geq 2$.

One rule sets solving inequalities apart from solving equations: *If both sides of an inequality are multiplied or divided by a negative number, the direction of the inequality must be flipped.*

Example 1: $-2x + 4 > 8$.

- First, we subtract 4 from both sides, obtaining $-2x > 4$.
- Next, we divide both sides by -2. However, once we do this, we have to change $>$ to $<$. Doing so, we find that $x < -2$.
- This means that all values of x that are less than -2 will satisfy the original inequality; as with most inequalities, this inequality has an infinite number of solutions.

Next, let us learn to solve systems of inequalities.

Example 2: $2x > 8$ and $3x > 7$. This system wouldn't be solvable as two equations, but it is solvable as inequalities.

- Solving both of these individually, we obtain $x > 4$ and $x > \dfrac{7}{3}$.
- Let us think about this logically. Since 4 is greater than $\dfrac{7}{3}$, if a number is greater than 4, it is automatically greater than $\dfrac{7}{3}$. Therefore, we can simplify the solution to just $x > 4$.

Example 3: $x > 45$ and $x < 90$.

- This can be read as "x is greater than 45 and x is less than 90." In this system, there is a set stretch of numbers that satisfies both conditions, as x can be greater than 45 and less than 90 at the same time.
- A way to denote the solutions to these types of systems of inequalities is the format $\_\_ < x < \_\_$. In this case, $45 < x < 90$. If this were what was given in the beginning, it could have been expanded back into the two inequalities $x > 45$ and $x < 90$.

Example 4: $x < 42$ and $x > 67$.

- x cannot be less than 42 and greater than 67 at the same time.
- Therefore, there is no solution.

Almost anything can be done to both sides of an equation. However, not everything is valid with inequalities.

For example, if $\dfrac{a}{x} > b$, it is not necessarily true that $a > bx$. If x were to be negative, the sign would have to be flipped.

Negatives are also an issue when raising both sides of an inequality to a power (both fractional and integer powers). For example, $-3 < -2$, but if we square both sides without flipping the inequality sign, we obtain $9 < 4$, which is untrue.

Many inequalities take a bit of logical thinking to solve.

Example 5: $x^2 > 75$.

- How do we solve this inequality for x? If x is greater than the positive square root of 75, x^2 will be greater than 75. However, there is another case that we must consider. What if x is less than the negative square root of 75? x^2 will be greater than 75 here as well, since the product of two negative numbers is positive.

- Therefore, the solution to this inequality is $x > \sqrt{75}$ or $x < -\sqrt{75}$.

An application of inequalities is comparing fractions.

Example 6: Which is greater, $\dfrac{3}{23}$ or $\dfrac{7}{52}$?

- Let us set up an inequality that represents this problem. How will we do this?

- If we put $\dfrac{3}{23}$ on the left side and $\dfrac{7}{52}$ on the right side, the operator in between can be $<$, $=$, or $>$. We will leave the operator as (?) for now.

- No matter what the (?) actually is, cross multiplying is valid. Therefore, from $\dfrac{3}{23}(?)\dfrac{7}{52}$, it is true that 3×52 (?) 23×7.

- $3 \times 52 = 156$ and $23 \times 7 = 161$. From this, we realize that the (?) is actually $<$. Going back to the original problem, $\dfrac{7}{52}$ is the greater fraction.

Problems: Inequalities

1 Bronze. Solve the inequality $a + 35 < 2a - 12$.

2 Bronze. Solve the inequality $-3x + 34 \geq 58$.

3 Bronze. Solve the inequality $x + 10 > 13$.

4 Silver. Solve the inequality $\dfrac{2x - 3}{-4} \leq 41$.

5 Bronze. For how many integer values of x is the inequality $3 < x \leq 12$ true?

6 Silver. To calculate the number of aliens spawned in *SpaceAlien-Wars2415*, you have to take your age, divide it by 3, and add 30. What is the minimum age that you must be to spawn 40 aliens?

7 Silver. $x + 3 > 45$ and $2x + 10 \geq 40$. Solve for x.

8 Silver. $2x + 31 > 32$ and $-3x - 45 > 5$. Solve for x.

9 Gold. $12x + 24 \leq 156$ and $-45x < 157.5$. Solve for x.

10 Platinum. Solve the inequality $35 + x < 43 - 2x \leq 5 - x$.

11 Gold. If $x^2 > 9$, what are the possible values for x?

12 Platinum. a is a positive constant greater than 5 and b is a negative constant. Solve the inequality $\dfrac{x^2}{b} > 1 - a$ for x.

13 Gold. $\sqrt{2x} \leq 3$. Solve this inequality for x.

14 Gold. Place the three fractions $\dfrac{4}{57}$, $\dfrac{2}{29}$, and $\dfrac{3}{43}$ in increasing order.

Part 8: Counting Numbers

Counting numbers is the first thing a child learns in the field of mathematics.

Example 1: Count all of the integers from 1 to 5.

- How many numbers did you count? You probably counted 5.

Here is a slightly harder problem:

Example 2: A person counts all of the integers from 103 to 569. How many numbers does the person count?

- One would think that the answer is $569 - 103 = 466$, but the answer is actually 467. The reason is that if the two numbers are subtracted, 103 is not accounted for. Therefore, 1 must be added to the result.

- However, 466 is the correct answer if 103 is not supposed to be included.

Example 3: I have already driven 13 miles with my car. How many more miles must I drive to get to 400 miles?

- Here, $400 - 13$ or 387 is the correct answer, as we are counting the number of *steps* from 13 to 400, not the numbers themselves.

You can derive all of these facts yourself by testing small intervals.

For example, counting all of the integers from 1 to 3 yields $3 - 1 + 1 = 3$ integers (count the integers in your head), so counting all of the integers from 99 to 132 will yield $132 - 99 + 1 = 34$ integers.

Problems: Counting Numbers

1 Bronze. How many integers are between 65 and 902 (not including 65 and 902)?

2 Bronze. If you count the integers from 445 to 33321, how many numbers do you count?

3 Silver. How many integers are greater than 1 but less than or equal to 339?

4 Silver. How many multiples of 8 are strictly between 0 and 99?

5 Gold. If it is now the beginning of December 3, 1989, in how many days will it be the beginning of June 9, 1990?

Part 9: Sequences and Series

Sequences (also known as progressions) are lists of numbers that follow a set pattern. For example, the list 2, 4, 6, 8, 10, ... is a sequence. The pattern is that each term is two plus the previous term. Another example of a sequence is 3, 12, 48, 192 .... Here, the pattern is that each term is 4 times the previous term.

If it is not given that a sequence has an end, it is assumed to be infinite. Two major types of sequences are arithmetic sequences and geometric sequences.

Arithmetic sequences are sequences where every 2 consecutive terms have the same difference.

For example, 4, 7, 10, 13, 16, 19... is an arithmetic sequence where the common difference is 3, and 2, 3, 4, 5... is an arithmetic sequence with a common difference of 1. The common difference is the difference between any two consecutive terms.

In any arithmetic sequence, if we call the first term a and the common difference d, the sequence can be written as a, $a + d, a + 2d, a + 3d, a + 4d, \ldots$. The n-th term is $a + d(n - 1)$.

Example 1: What is the 24th term of the sequence $12, 15, 18, \ldots$?

- In this sequence, the starting term is 12 and the common difference is 3.
- We can write this sequence as $12, 12 + 3, 12 + 2(3), 12 + 3(3)\ldots$, where the n-th term is $12 + 3(n - 1)$. The 24th term of the sequence is therefore $12 + 3(23)$ or 81.

Geometric sequences are sequences where every two consecutive terms have the same quotient. For example, $2, 8, 32, 128, \ldots$ is a geometric sequence where the common quotient, or the quotient of any term in this sequence and the term before it, is 4.

In any geometric sequence, if we call the first term a and the common quotient q, the sequence can be written as $a, aq, aq^2, aq^3, aq^4 \ldots$. The n-th term can be found using the formula $a \times q^{(n-1)}$.

Example 2: What is the eighth term of the sequence $1, 2, 4, 8 \ldots$?

- In this sequence, the starting term is 1 and the common quotient is 2.
- We can write this sequence as $1, 1 \times 2, 1 \times 2^2, 1 \times 2^3 \ldots$, where the n-th term is $1 \times 2^{n-1}$. The eighth term of the sequence is therefore 1×2^7 or 128.

The sum of the terms in a sequence is called a series. Arithmetic series are the sum of the terms in an arithmetic sequence, and vice versa.

In any arithmetic series, the sum of the first and last terms is the same as the sum of the second and the second to last, which is also the same as the sum of the third and third to last, and etc.

If the series has n terms, the number of these pairs is $\dfrac{n}{2}$ (If the series has an odd number of terms, the middle term comes out every time to be exactly half of a pair, so this still works).

Knowing this, we can derive that the formula for arithmetic series is $\dfrac{n(a + f)}{2}$, where a is the first term of the sequence, f is the last term, and n is the number of terms.

Special cases can be constructed from this formula: One can use it to derive that the sum of the first n positive integers $(1+2+3\ldots+n)$ equals $\dfrac{n(n + 1)}{2}$, the sum of the first n odd positive integers is n^2, and the sum of the first n even positive integers is $n(n + 1)$

Example 3: Consider the arithmetic sequence $3, 6, 9, 12, 15\ldots$. Find the sum of the first 20 terms in this sequence.

- The first term of the sequence is 3, and the common difference is also 3. The last term of the sequence is $3 + 19(3)$ or 60.

- The sum of the first and last terms is 63. The sum of the second and second-to-last terms is also 63. The same goes for the third term plus the third-to-last term, the fourth term plus the fourth-to-last term, etc.

- This sequence has 20 terms, so there are 10 of these pairs. Therefore, the sum of the first 20 terms in the sequence is $63(10)$ or 630.

The sum of a geometric series $= a\left(\dfrac{1 - q^n}{1 - q}\right)$, where a is the first term in the sequence, n is the number of terms, and q is the common quotient. We will show why this is true later in the chapter.

There exist other sequences besides those that are arithmetic and geometric. However, all sequences have some sort of pattern. For example, one can easily identify the pattern in the sequence $1, 11, 111, 1111, \ldots$ even though it is neither arithmetic nor geometric.

In a sequence such as $\dfrac{1}{2}, \dfrac{3}{4}, \dfrac{5}{8}, \dfrac{7}{16}, \dfrac{9}{32}$, the numerators form an arithmetic sequence while the denominators form a geometric sequence.

Problems: Sequences and Series

1 Bronze. What is the sum of the first 15 positive integers?

2 Bronze. What is the sum of the first 32 positive integers?

3 Silver. What is the sum of the series $3, 7, 11, 15, \ldots, 43, 47$?

4 Bronze. The first term of an arithmetic sequence is 11 and the common difference is 4. What is the 16th term?

5 Bronze. What is the 25th term of the sequence $-7, -3, 1, 5, 9, 13, 17, 21, \ldots$.

6 Bronze. What is the sum of the first 22 odd positive integers?

7 Bronze. If the first term in an increasing arithmetic sequence is 3 and the difference between consecutive terms is 6, what is the 10th term?

8 Gold. If the third term in an arithmetic sequence is 29 and the eight term is 38, what is the 15th term? Express your answer as a decimal to the nearest tenth.

9 Bronze. What is the fifth term of a geometric progression with first term 2 and second term 3?

10 Bronze. Find the seventh term of the sequence $3, 6, 12, 24, \ldots$.

11 Silver. The sum of 5 consecutive odd integers is 115. Find the product of the greatest two of these integers.

12 Silver. There are 20 teams in a soccer league. Each team plays every other team exactly once. How many total games are played?

13 Bronze. Find the value of the 47-term series $1 - 3 - 7 - 11 - 15 - 19 - 23 - 27 - 31 \ldots - 179 - 183$.

14 Gold. Consider the sequence $\dfrac{15}{2}, \dfrac{20}{3}, \dfrac{25}{4}, \dfrac{30}{5}, \dfrac{35}{6}, \ldots$. Find an expression to determine the n-th term of this sequence in terms of n.

15 Silver. $\dfrac{x}{1}, \dfrac{2x}{3}, \dfrac{3x}{5}, \dfrac{4x}{7}, \dfrac{5x}{9} \ldots$ What is the 502nd term of this sequence?

16 Gold. Find the 100th term of the sequence 1, 3, 6, 10, 15, ...

Part 10: Distance, Rate, and Time

For all problems involving distance, rate, and time, there is one formula that you must remember: Distance = Rate × Time

Rate (a.k.a. speed in these type of problems) is the amount of distance traveled in one unit of time, or how fast something is. It is expressed as a unit of distance over a unit of time. Examples of units for rate are miles per hour and meters per second.

Example 1: Yiem is riding his bike at an average speed of 25 miles per hour for 3 hours. How far does he travel?

- To solve this problem, we use the formula *Distance = Rate × Time*.

- The distance is what we are trying to find, but we already have the rate and the time, which are 25 and 3 respectively. We plug these values into the equation, getting 75 miles as our answer.

Example 2: Yiem walks along a straight road at a constant speed for 3 hours, and then he jogs the remaining 4 miles before reaching the road's end. If he had kept his walking pace, Yiem could have traveled the same distance in 5 hours. Find Yiem's walking pace.

- Let us call Yiem's walking pace in miles per hour y. How far does he travel in 3 hours?

- Since *Distance = Rate × Time*, Yiem travels $3y$ miles in 3 hours. After walking the $3y$ miles, Yiem jogs another 4 miles. The total distance he travels is $3y + 4$ miles.

- With the same pace y, Yiem can travel the same $3y + 4$ miles in 5 hours. According to the formula *Distance = Rate × Time*, traveling at y miles per hour for 5 hours yields a distance of $5y$ miles.
- We know that the $5y$ miles is the same distance as the $3y + 4$ miles, so $5y = 3y + 4$. Solving this equation, we find that Yiem's walking pace is 2 miles per hour.

(*Note*: Make sure your units remain constant throughout the problem, and make sure your unit for rate consists of the same units as your distance and time.)

Problems: Distance, Rate, and Time

1 Bronze. How many miles will a cyclist travel in 3 hours if he is moving at 30 miles per hour?

2 Bronze. If a car travels 50 miles in 2 hours, what is its average speed in miles per hour?

3 Silver. How long does it take a car moving at 12.2 miles per hour to travel 42.3 miles? Round your answer to the nearest tenth.

4 Silver. One runner's pace is 30 yards per minute. Another runner's pace is 50 yards per minute. If these two runners race against each other and the slower runner gets a six-minute head start, for how much time will the faster runner have run when he catches up with the slower runner?

5 Silver. A runner is traveling at a speed of 25 meters per minute around the perimeter of a regular octagonal building with side length 30 meters. If he continuously runs from 8:00 to 8:30 a.m., how far does he travel in meters?

6 Platinum. A train moves at 40 miles per hour towards City A from the train station. After $1\frac{1}{2}$ hours, it stops for 45 minutes due to engine trouble. Once its engine is repaired, it travels at 50 miles per hour for the remainder of the trip, and is exactly on time. If

the train had traveled at 40 miles per hour the whole trip and never stopped, it would have also been exactly on time. How far is it from the train station to City A?

7 Gold. Two runners are jogging around a circular track that is 100 feet long. One runner is three times as fast as the other. If they start at the same point on the track and go in opposite directions, how far will the slower runner have run when they meet up for the first time after they start?

8 Silver. Two runners jog around a circular track 100 feet long in the same direction. The first runner jogs at 10 feet per minute, and the second runner jogs at 12 feet per minute. If they start at the same time and place, after how much time will the faster runner lap the slower one?

9 Gold. Two recreation centers are 50 feet apart. A man jogs from the first recreation center to the second at a pace of 5 feet per second, and returns from the second recreation center to the first at a pace of 10 feet per second. Find his average pace in feet per second. Express your answer as a common fraction.

Part 11: Rates

Most problems that involve rates can be placed under one category. Our main formula for these problems is *rate × time = amount of stuff done*. Our formula for distance, rate, and time is a special case of this formula.

In these types of problems, the rate is sometimes tricky to find. To find the rate of something, you have to find the amount it does in one unit of time. For example, if a person completes a job in 3 hours, then he or she does $\frac{1}{3}$ of a job in 1 hour, so his or her rate is $\frac{1}{3}/1$, which is just $\frac{1}{3}$ (in jobs per hour).

Example 1: Bob can mow his lawn in 3 hours. If he works at a constant rate, how long will it take him to mow two lawns both the same size as his?

- Bob mows one lawn in 3 hours, so his rate is $\frac{1}{3}$ lawns per hour. The amount of work he is going to be doing is 2, since he will be mowing two lawns.

- Let us call the time he takes t, since that is what we are trying to find.

- Now we set up the equation $t \times \frac{1}{3} = 2$. Solving this equation, we find that $t = 6$, so it takes 6 hours for Bob to mow the two lawns.

Rates, times, and amounts of "stuff" can each be added and subtracted. In the case of rates, if two rates are working together, they can be added, and if two rates are counteracting each other, then they can be subtracted.

Example 2: Bob can mow his lawn in 4 hours and Jim can mow the same lawn in 3 hours. Working together, how long will it take them to mow 4 lawns of the same size?

- Bob's rate is $\frac{1}{4}$ lawns per hour, and Jim's rate is $\frac{1}{3}$ lawns per hour. When they work together, their combined rate is $\frac{1}{4} + \frac{1}{3}$ or $\frac{7}{12}$.

- Using our formula, we set up the equation $t\left(\frac{7}{12}\right) = 4$, solving for the time t. Therefore, $t = \frac{48}{7}$ hours.

(*Note:* As in the previous section, make sure your units stay constant throughout the problem.)

Example 3: What is the first time after 4:00 p.m. at which the minute hand and hour hand of a clock will meet up rounded to the nearest minute?

- In this problem, we will calculate the rates of the minute and hour hands in revolutions per minute.

- Our main formula will be rate (in revolutions per minute) × time (in minutes) = revolutions. This is identical to rate × time = amount of work done, except it is a special case.

- Let us picture our desired scenario. This scenario, where the minute and hour hands coincide, will be sometime after 4:20 but before 4:25.

- Let us now find the rates of the minute and hour hands. Since the minute hand of a clock makes one full revolution every 60 minutes, it makes $\frac{1}{60}$ revolutions per minute. Since the hour hand of a clock makes one full revolution every 12 hours or 720 minutes, it makes $\frac{1}{720}$ revolutions per minute.

- At 4:00 p.m., the hour hand has a $\frac{1}{3}$ of a revolution head start on the minute hand. We want the fraction of a revolution that the minute hand makes to equal the fraction of a revolution that the hour hand makes plus the head start that it has.

- If we set the time that this takes as t, we can write the equation $\frac{1}{60}(t) = \frac{1}{720}(t) + \frac{1}{3}$. Solving this equation, we find that $t = \frac{240}{11}$ or $21\frac{9}{11}$ minutes. Our answer is this amount of time after 4:00 p.m., which turns out to be 4:22 to the nearest minute.

Problems: Rates

1 Silver. One hose can fill a bottle in 3 minutes. Another hose can fill the same bottle in 2 minutes. If they work together to fill the bottle, how long will it take in minutes? Express your answer as a decimal to the nearest tenth.

2 Silver. One machine can assemble a toy in 32 hours. Another machine can assemble the same toy in 48 hours. Working together to assemble the toy, how long will the machines take? Express your answer as a decimal to the nearest tenth.

3 Gold. One man can shovel a road alone in 10 hours. He works with his son and they complete the road in 3 hours. Assuming both of them work at a constant rate, how long will the son take to complete the road on his own? Express your answer as a common fraction.

4 Gold. Brock and Sei are sharing a pizza. Brock eats some of the pizza in two hours and gives it to Sei, who finishes eating it in an hour. If Sei had started eating the pizza himself for four hours, then Brock would have been able to finish what was left in one hour. Assuming both people eat at a constant rate, how many hours would it have taken for Brock to eat the pizza alone?

5 Silver. A tub can hold 30 gallon of water. Vode is going to fill it up with faucets that each give out water at a rate of 0.6 gallons per hour. How many faucets must he use if he needs to fill the tub completely in 2 hours?

6 Gold. There are 55 seats in a room. Hamie can clean four seats in a minute, and Erok can clean three in a minute. Hamie cleans seats alone for five minutes, and then Erok joins him until they finish cleaning all of the seats in the room. For how many minutes does Hamie clean seats in total?

7 Silver. One motor engine spins a wheel clockwise at a rate of 32 revolutions per second. Another motor spins the wheel counter-clockwise at a rate of 54 revolutions per second. If both of these motors act on the same wheel simultaneously, how many revolutions will the wheel make in 10 seconds?

8 Silver. Two teams of people are playing Tug of War. One team is pulling the rope west at a rate of 20 meters per second. The other team is pulling east at 22 meters per second. The team that moves the rope 10 meters in the direction they are pulling wins. How long will it take for one of the teams to win?

9 Gold. It is now 4:00 p.m. What is the earliest time at which the sum of the minute hand's and hour hand's movements add up to a full revolution? Round your answer to the nearest minute.

10 Gold. A 4A battery will last 3 hours in a flashlight before dying, 6 hours in a fan before dying and 12 hours in a speaker before dying. One 4A battery is put in a flashlight for half an hour, and then in a fan for two and a half hours. If it is now put in a speaker, how long will it last before dying?

11 Platinum. What is the earliest time at which the minute and hour hands of a regular analog clock meet after 12:00 a.m.? Express your answer to the nearest second.

12 Gold. One person can clean 1 seat in 6 minutes. Another person can clean 6 seats in one minute. Working together, how long will it take them to clean the 74 seats in an auditorium?

13 Gold. It is now 4:15. The minute hand of a regular analog clock is on the 3, and the hour hand is past the 4. When will the minute hand catch up to the hour hand? Round your answer to the nearest second.

14 Silver. The variable y increases by 3 per increase of 1 by the variable x. If y increases by 42, by how much does x increase in return?

Part 12: Unit Conversion and Analysis

One application of proportions is conversion between units.

Example 1: Convert 35 hours to minutes.

- Let us define m as the number of minutes in 35 hours. The proportion $\dfrac{35\,\text{hours}}{m\,\text{minutes}}$ equals 1, since these two amounts, although they have different units, are equal.

- We know that there are 60 minutes in an hour. Therefore, the proportion $\dfrac{1\,\text{hour}}{60\,\text{minutes}}$ also equals 1.

- Knowing this, we set up the equation $\dfrac{35\,\text{hours}}{m\,\text{minutes}} = \dfrac{1\,\text{hour}}{60\,\text{minutes}}$.

When solving a proportional equation with two or more different units, make sure that the units in the numerators and the units in the denominators match up on both sides of the equation.

For example, $\dfrac{35\,\text{hours}}{m\,\text{minutes}} = \dfrac{60\,\text{minutes}}{1\,\text{hour}}$ would not be easy to solve even though the net value of both sides is 1. This is because the unit minutes is in the numerator of the right side but in the denominator of the left side, and the unit hours is in the numerator of the left side but in the denominator of the right side.

If we match up the units correctly, we can temporarily disregard them and solve for m. Doing this, we find that $m = 2100$ minutes.

In mathematics and in other fields, we are often able to use units to help us remember formulas. The way to do this is to treat the units like we treat variables.

Consider *Distance = Rate × Time*. Let the unit for distance be miles, the unit for time be hours, and the unit for rate be miles per hour. Note that any unit "x per y" can be written as $\dfrac{x}{y}$.

We know that for any values of a and b, $a \times \dfrac{b}{a} = b$. We can use this thinking process with units. Say that $x\,\dfrac{\text{miles}}{\text{hr}} \times y\,\text{hours} = z\,\text{miles}$. Notice how the two occurrences of hours cancel each other out if we treat units as variables.

We can also use this same method to recall that *Distance/Rate = Time* and *Distance/Time = Rate*.

Say that a wheel rotates at $32\,\dfrac{\text{revolutions}}{\text{minute}}$ for 10 minutes. If we multiply $32\,\dfrac{\text{revolutions}}{\text{minute}}$ and 10 minutes, the two occurrences of minutes cancel, and we are left with just the unit revolutions. $32 \times 10 = 320$.

This analysis also works for single unit expressions. For example, 3 hours + 5 hours = 8 hours just as $3x + 5x = 8x$.

Example 2: Given that pressure is expressed in Newtons/m², Newtons are units of force, and m² are units of area, find a formula solving for pressure.

- Since the unit for pressure is Newtons/m², if we divide a quantity in Newtons by a quantity in meters², treating units as variables should leave the final answer in the unit Newtons/m².

- Newtons is a unit for force and m² is a unit for area, so Pressure = Force/Area.

Only use simple addition, subtraction, multiplication, and division with this type of unit analysis. The analogy of treating units as variables only works to a certain extent. For example, even though acceleration is expressed in meters per second squared, acceleration does not equal distance divided by time squared. For the problems in this section, however, you can assume that this type of unit analysis is always correct.

Problems: Unit Conversion and Analysis

1 Silver. Gums, roqs, and yams are units of currency on a distant planet. 2125 gums = 425 yams = 1 roq. How many gums are equivalent to 13 yams?

2 Silver. There are 1760 yards in a mile and 3 feet in a yard. A racetrack in Sedona is 1 mile, 2 yards and 2.4 feet long. How long is the racetrack in yards?

3 Bronze. The mileage of a certain car is 25 miles per gallon. How many gallons will be used on a trip that is 200 miles long?

4 Gold. Specific heat capacity, the amount of heat energy needed to raise one gram of a material by 1 degree Celsius, is expressed with the unit $\dfrac{\text{Joules}}{\text{grams} \times {}^\circ\text{Celsius}}$. How many Joules would it take to raise 10 grams of a material with a specific heat capacity of 8 $\dfrac{\text{Joules}}{\text{grams} \times {}^\circ \text{Celsius}}$ by 10 degrees Celsius?

5 Silver. A car travels at 20 feet per second. Express the car's speed as a common fraction in miles per hour. Note that there are 5,280 feet in a mile.

Part 13: Percentages

Percentages are another way of writing fractions. Something percent basically means that something over 100. For example, $30\% = \dfrac{30}{100}$, which can be simplified to $\dfrac{3}{10}$. Also, 33.2% equals $\dfrac{33.2}{100}$, which can be written in decimal form as 0.332. If both numerator and denominator are multiplied by 10 and then the resulting fraction is simplified, 33.2% can be converted into the fraction $\dfrac{83}{250}$.

Remember that multiplying both the numerator and the denominator of a fraction by the same number is equivalent to multiplying the entire fraction by 1. Therefore, doing this does not change the fraction's value.

Percentages can be converted to fractions using this method, and fractions can be converted to percentages with a simple proportion.

Example 1: Convert $\dfrac{1}{2}$ to a percent.

- We define the percent number we are looking for as x.
- $\dfrac{x}{100}$ will equal the fraction we are converting by the definition of a percent, so we can set up the equation $\dfrac{x}{100} = \dfrac{1}{2}$.
- Solving, we find that $x = 50$, so $\dfrac{1}{2} = 50\%$.

Percent of change is an important concept involving percentages. It is used often in the real world, such as when you are at a store and an item is "40% off," or when an advertisement for the new model of a cell phone states that the battery life is "60% longer than before."

The percent of change formula is this: percent of change $=$ $\dfrac{\text{amount of change}}{\text{original amount}} \times 100$.

Example 2: What is the percent decrease from \$48 to \$36?

- The amount of change from 48 to 36 is 12, and the original amount is 48. Therefore, the percent of change is $\left(\dfrac{12}{48} \times 100\right)$%, or 25%.

We can also reverse this method of calculation:

Example 3: What number represents a 40% increase from 35?

- If we call the amount of change x, we can set up the equation $40 = \dfrac{x}{35} \times 100$. Solving, we find that $x = 14$. Since the amount of change is 14, the answer is $35 + 14 = 49$.

Here is a shortcut for percent of increase and decrease: increasing a number by x% has the same effect as multiplying it by $\left(1 + \dfrac{x}{100}\right)$, and decreasing a number by x% has the same effect as multiplying it by $\left(1 - \dfrac{x}{100}\right)$. Using this method, increasing 35 by 40% equals $35 \times (1 + 0.4)$, or 49.

An important application of percentages is calculating interest. When money is borrowed, interest is the amount of money paid to the lender in addition to the borrowed amount.

There are two types of interest: simple and compound.

Simple interest is a percentage of the borrowed amount added every certain period of time. This percentage always acts on the original amount of borrowed money.

Compound interest builds up on itself. Each time that compound interest is taken, the interest acts on the original amount plus all of the previous interests added to it.

For example, 5% simple interest taken yearly on a loan of \$100 over four years yields 4 times 5% of 100 dollars, which equals $4\left(\dfrac{5}{100} \times 100\right)$ or \$20.

5% compound interest taken yearly on a loan of \$100 over four years equals $\dfrac{5}{100} \times 100$ or \$5 PLUS $\dfrac{5}{100} \times (100 + 5)$ or \$5.25 PLUS $\dfrac{5}{100} \times (100 + 5 + 5.25)$ or \$5.51 to the nearest cent PLUS $\dfrac{5}{100} \times (100 + 5 + 5.25 + 5.51)$ or \$5.79. All of this added up equals \$21.55.

Example 4: Ino deposits $200 into a bank account. At the end of each year, the bank increases the amount of money in the account by 1%. How much money will be in the account at the end of three years?

- The bank provides compound interest on Ino's deposit, since the amount of interest on any given year depends on the amount of money currently in the account, not on the size of the original deposit.

- 1% of $200 is $2. At the end of one year, the account will hold $200 + $2 or $202.

- 1% of $202 is $(0.01 \times \$202) = \$2.02$. At the end of two years, the account will hold $202 + $2.02 = $204.02.

- 1% of $204.02 is $(0.01 \times \$204.02) = \$2.04$ to the nearest cent (always round to the nearest cent when dealing with money). At the end of three years, the account will hold $204.02 + $2.04 = $206.06.

Problems: Percentages

1 Bronze. Write the fraction $\frac{4}{5}$ as a percent.

2 Bronze. Write the fraction $\frac{3}{8}$ as a percent.

3 Bronze. Write 45% as a fraction.

4 Bronze. What is 50% of 320?

5 Silver. One day, there were 50 bacteria in a jar. The next day, there were 75 bacteria in the jar. What percent increase does this represent?

6 Gold. There are c children in a room and o cookies. If the cookies are equally distributed among the children, what percent of the cookies does each child get in terms of c and o?

7 (calculator) Gold. Every year, the number of ants in a tree increase by 30%. After 3 years, by what percentage would have the amount of ants in the tree increased in total?

8 Gold. A man is buying a scarf. He uses three 20% off coupons, each of which acts on the updated price, not the original

price. What single percent discount represents the discount he receives using his three coupons?

9 Silver. Find 32% of $4xy$ in terms of x and y.

10 Silver. A rectangle is 25 meters wide. This width is going to be multiplied by $\frac{7}{5}$. What percent increase does this represent?

11 Silver. A company put $4000 in a savings account. This account pays simple interest of 10% each year. How much profit will the company gain after four years?

12 Gold. A family takes a loan for $40 over three months. The interest rate is 5%, and is compounded every month. How much money will the family have to pay back in total?

Part 14: Mixtures

Mixture problems are common word problems that involve substances being mixed together. These problems usually involve a combination of amounts and percentages, which is what makes them difficult.

To solve mixture problems, we have to first convert the percentages to amounts.

For example, if there are 5 gallon of a juice that is 10% sugar and the rest water, then the juice contains 0.5 gallon of sugar and 4.5 gallon of water. Knowing these amounts, it is much easier to set up an equation and eventually solve the problem.

Usually, the equation you set up should be a proportional equation where the numerator is the amount of ingredient and the denominator is the total amount of the substance. This equals your desired fraction of ingredient out of the total.

Example 1: Bob has 20 liters of a mixture that is 50% soda. He has 200 liters of a mixture that is 75% soda. How much of the second mixture must he add to the first to make it 65% soda?

First, we convert the percentages in the problem to amounts. The first solution is 50% soda, and 50% of the total 20 mL is 10 mL.

Therefore, the first mixture contains 10 mL of soda. We do not know how much of the second mixture we are going to be using, so let us call the amount x.

Since it is a 75% soda mixture, $\frac{3}{4}x$ is soda. We need the solution to be 65% or $\frac{13}{20}$ soda; that will be the right side of our equation. Now, let us set it up. The total amount of soda in the mixture is the numerator, which is $10 + \frac{3}{4}x$.

The denominator is the total amount of mixture, which is $20 + x$. The problem requires that $\left(10 + \frac{3}{4}x\right) / (20 + x) = \frac{13}{20}$. Solving this equation, we find that $x = 30$ mL.

Problems: Mixtures

1 Silver. A scientist has 5 mL of a solution that is 50% acid. How much pure water must he add to make the solution 25% acid?

2 Gold. A scientist has 30 mL of a solution that is 20% acid and the rest water. He has another solution that is 10% water and the rest acid. How much of the second solution must he add to the first solution to make it 30% acid?

3 Silver. A scientist has 20 mL of a solution that is 50% acid. How much pure water must be added to the solution to make it 10% acid?

4 Gold. A jar contains red, blue, green, and yellow cookies. The jar is currently 10% red cookies, 25% blue cookies, 35% green cookies, and 30% yellow cookies. Another jar of 70% yellow cookies and 30% blue cookies contains half as many cookies as the first jar. If the second jar's contents are dumped into the first, what will be the overall percentage of yellow cookies? Round your answer to the nearest tenth of a percent.

5 Silver. A man has a pot containing 40 gallons of a liquid that is 25% mud. The mixture has to be 20% mud to be safe to drink for his animals. How much pure water must he add to the mixture to make it safe to drink?

Part 15: Polynomial Expansions

Let us examine the expression $(2a - 3b)$. This expression is classified as a binomial. Terms such as 3 or a are monomials, and expressions such as $(a + b + c)$ are trinomials.

To figure out what kind of "nomial" your expression is, count the number of terms added or subtracted in the expression when it is in simplest form. Polynomials consist of all "nomials" except monomials. Binomials and trinomials are types of polynomials.

The degree of a term is the sum of the exponents of its variables. For example, the degree of the term a^6 is 6, the degree of the term ab is $1 + 1$ or 2, and the degree of the term $7x^2y^3$ is $2 + 3$ or 5. The degree of a polynomial is the highest degree out of any of its terms. For example, the degree of the polynomial $x^6 + x^4y^3 + xy$ is 7, as 7 is the highest degree out of any of the polynomial's terms.

Similar variables with different exponents cannot be combined through addition or subtraction. Therefore, an expression such as $b^2 + b + 1$ is in simplest form. The previous expression is a trinomial, but it only contains one unique variable.

When binomials, trinomials, and polynomials are multiplied with other binomials, trinomials, and polynomials, the result can be simplified into one polynomial through a process called expansion.

Recall the distributive property. It says that expressions such as $x(y + z)$ can be simplified into $xy + xz$. This is the way to expand the product of a monomial and a polynomial.

Example 1: Expand $x(2x + 3xy)$.

- By the Distributive Property, $x(2x+3xy) = (x \times 2x)+(x \times 3xy)$.
- $(x \times 2x) = 2x^2$, and $x \times 3xy = 3 \times x \times x \times y = 3x^2y$.
- Therefore, $x(2x + 3xy) = 2x^2 + 3x^2y$.

To expand the product of two polynomials, multiply every term in the first polynomial by every term in the second. Then, add all of the resulting products.

Example 2: Expand $(a + b)(a - b)$.

- First, we multiply the two a's, making a^2.
- Next, we multiply the first a by the $-b$, getting $-ab$.
- Then we multiply the b in the first binomial by the a in the second, making a positive ba, which can be reordered into ab to match up with our other terms (Commutative Property of Multiplication).
- Lastly, we multiply the b and $-b$ to make $-b^2$. Now we add all of the resulting terms, obtaining $a^2 + ab - ab - b^2$.
- The ab and $-ab$ cancel, so our final answer is $a^2 - b^2$.

Note that when expanding, we treat subtraction in the polynomials as negative signs attached to the terms in front of the subtraction signs. We treat a term like $(a - b)$ as $(a + (-b))$, since otherwise we won't know how to deal with the subtraction.

Example 3: Expand $(2y + 3)(3x + y - 1)$.

There are six products that we must find and add.

- $2y \times 3x = 6xy$
- $2y \times y = 2y^2$
- $2y \times (-1) = -2y$
- $3 \times 3x = 9x$
- $3 \times y = 3y$
- $3 \times (-1) = -3$
- The final answer is $9x + 6xy + 2y^2 + y - 3$.

It would be useful to simply memorize a few common expansions.

1. $(a + b)^2$ [also known as $(a + b)(a + b)$] $= a^2 + 2ab + b^2$
   - $(a + b)(a + b) = (a \times a) + (a \times b) + (b \times a) + (b \times b)$
     $$= a^2 + ab + ab + b^2 = a^2 + 2ab + b^2.$$
2. $(a + b)(a - b) = a^2 - b^2$.

We previously went through this expansion as an example.

Standard order is the order in which polynomials generally should be put. To put a polynomial in standard order, you take the variable that is lowest in the alphabet and put the polynomial in order from highest to lowest power of that variable.

Smaller expressions, such as $(a + b)^2$, are easy to expand, but what about expressions like $(a + b)^6$? This would take quite a bit of time. Luckily, there is a tool called Pascal's triangle to help us with expansions like this.

$$
\begin{array}{c}
1 \\
1 \ 1 \\
1 \ 2 \ 1 \\
1 \ 3 \ 3 \ 1 \\
1 \ 4 \ 6 \ 4 \ 1 \\
1 \ 5 \ 10 \ 10 \ 5 \ 1 \\
1 \ 6 \ 15 \ 20 \ 15 \ 6 \ 1
\end{array}
$$

Every row starts and ends with a 1, and every number in between is the sum of the two numbers directly above it (look at the figure and then try to recreate it without looking—you can start at the top and then easily figure out where each number fits). All rows are symmetrical. One quirk about this triangle is that the top row of the triangle is labeled as row 0, not the first row.

The n-th row of the triangle represents the coefficients of $(a + b)^n$ in standard order. Each term in the polynomial has the powers of its two variables adding up to n.

Example 4: Expand $(a + b)^6$.

- $(a+b)^6 = a^6 + 6a^5b + 15a^4b^2 + 20a^3b^3 + 15a^2b^4 + 6ab^5 + b^6$, since the sixth row of Pascal's triangle is 1 6 15 20 15 6 1.

- Notice how we start at a^6b^0 (the b^0 simplifies to 1 and does not need to be written), and each time another term is added we decrease the power of a by 1 and increase the power of b by 1 until we hit a^0b^6.

Problems: Expansions

1 Gold. If $a + \dfrac{1}{a} = 12$, then what is the value of $a^2 + \dfrac{1}{a^2}$?

2 Bronze. Expand $a(b + 3)$.

3 Silver. Expand $(2n + 3)(n + 1)(3n + 2)$.

4 Silver. What is the largest coefficient of any term in the expression $(m + n)^6$?

5 Bronze. Is $(a^2 + a + a)$ a monomial, binomial, trinomial, or other polynomial?

6 Bronze. Expand $(a + b)^2$.

7 Silver. Expand $(m + n)^4$.

8 Silver. Find the value of $29^2 + 22 \times 29 + 11^2$ without writing anything down.

9 Bronze. In terms of a and b, what is $(a + b)^2 - (a - b)^2$.

10 Silver. Expand $(ab + cd)(ef + gh)$.

11 Bronze. Find the value of $(13 \times 14)\left(\dfrac{2}{13} + \dfrac{3}{14}\right)$.

12 Silver. Expand $(x + 2y)^3$.

13 Gold. If a company puts d dollars into a savings account with compound interest of $p\%$ for three years applied annually, find how much money the savings account will be worth after three years in terms of d and p.

14 Platinum. Expand $(3xy + 4)^5$.

15 Bronze. Expand $(3x + 2y + 4z^2)(x + z)$.

Part 16: Equivalent Expressions

The system of equations $2x + 3 = 4$ and $2x + 3 = 7$ has no solution, as $2x + 3$ and $2x + 3$ are equivalent expressions. Trying to solve this system through elimination or substitution yields a universal untruth.

What if we are given that two expressions are equivalent for all values of x, and are asked to solve for variables within the expressions?

Example 1: If the expression $2x + 4$ is equivalent to the expression $cx + c + d$ for all values of x, what are the values of c and d?

- Since we are solving for an expression in terms of x, we must leave x alone. x cannot hold any value or become anything other than x.
- Therefore, cx must match up with $2x$ if the two expressions are equivalent. It follows that $c = 2$.
- We now have $2x + 2 + d = 2x + 4$. It is easy to see that $d = 2$.

Problems: Equivalent Expressions

1 Silver. If the polynomial $ax^2 + bx^2 + ax - bx + c$ is equivalent to the polynomial $7x^2 + 4x + 8$, what are the values of a, b, and c?

2 Silver. The expression $2ax + 3bx + 3a + 2b$ is equivalent to $22x + 34$. Find the value of $a + b$.

Part 17: Factoring

Factoring is the opposite of expansion. When factoring, we take one polynomial and split it up into the product of different monomials and polynomials.

To factor a polynomial, you have to find its greatest common factor. This is the largest possible term that divides every term in the polynomial without any remainder. For example, the greatest common factor of $(2x + 4)$ is 2, because 2 is the greatest term that divides both $2x$ and 4 evenly.

Example 1: Factor the expression $2x^2 + 4x^6 + 8x$.

- $2x$ divides all the terms in the polynomial without any remainder. (Remember not to forget factors that are variables.)

- Now we divide all the terms in the polynomial by $2x$, obtaining $x + 2x^5 + 4$. Lastly, we multiply the factor by the factored polynomial, obtaining $2x(x + 2x^5 + 4)$.
- $x + 2x^5 + 4$ has no common factor, so we are done.

It is very important to learn how to factor trinomials in the form $ax^2 + bx + c$, where a, b, and c are whole numbered constants.

These trinomials can be factored into two binomials, $(ex + f)$ and $(gx + h)$ for some values of e, f, g, and h.

- If we expand these two binomials, we obtain $egx^2 + ehx + fgx + fh$.
- In order for this to coincide with $ax^2 + bx + c$ based on the powers of x, eg has to equal a, $eh + fg$ has to equal b, and fh has to equal c.
- Therefore, *if there exists values of e, f, g, and h where e* $\times$ *g = a*, *eh + fg = b*, and *f* $\times$ *h = c*, then $ax^2 + bx + c$ can be factored into $(ex + f)(gx + h)$.

Unfortunately, the quickest way to find the values of e, f, g, and h is to simply guess and check.

Not all trinomials in the form $ax^2 + bx + c$ can be factored in this way into nice numbers. It will not always be possible to find integer values for e, f, g, and h, but you will learn later that any trinomial can be factored in this way with either real or *imaginary* numbers.

Example 2: Factor $2x^2 + 9x + 9$.

- The only pair of whole numbers that multiplies to form 2 is 1 and 2, so we will set $e = 1$ and $g = 2$ (If we treat $e = 2$ and $g = 1$ as a separate case, we will receive the same pairs of binomials, except in opposite order. However, we cannot discount reversed orders in both e and g and f and h; we can only choose one of these pairs).
- There are two pairs of whole numbers that multiply to 9: 9, 1 and 3, 3. If we set f as 1 and h as 9, $eh + fg = 11$, so this does not work. If we flip it and set f as 9 and h as 1, $eh + fg = 19$, so

this does not work either. However, if we set f as 3 and h as 3, $eh + fg$ is in fact 9.

- Plugging the values of these variables into $(ex + f)(gx + h)$, we receive our answer, $(x + 3)(2x + 3)$.

Example 3: Factor $4x^2 + 4x + 1$.

- This trinomial will be factored into the form $(ex + f)(gx + h)$. The only two whole numbers that multiply to 1 are 1 and 1, so both f and h must equal 1 if this trinomial is factorable.

- There are two sets of numbers that multiply to 4: 1 and 4, and 2 and 2. Therefore, e and g must either be 1 and 4 in either order or 2 and 2.

- If e and g are 1 and 4, $eh + fg$ comes out to be 5, but we want it to be 4. However, when e and g are both 2, $eh + fg$ does equal 4.

- Therefore, the final factorization is $(2x + 1)(2x + 1)$, which can also be written as $(2x + 1)^2$.

Remember that these variables are only placeholders; they are meant only as an aid to help you work visually or intuitively.

Factoring can also be used to divide polynomials. Any real number or expression divided by itself is 1, including polynomials. If you have the expression $\dfrac{(x + 3)(x + 5)}{x + 3}$, $(x + 3)$ can be canceled out on both top and bottom, leaving only $(x + 5)$. However, when $x = -3$, $(x + 3) = 0$, so the original expression involves division by 0. This causes many complications, so we will write the solution as $(x + 5)$ for all x except -3.

If given the expanded version of $(x + 3)(x + 5)$, $x^2 + 8x + 15$, use factoring to determine which binomial can be canceled.

The method of factoring we discussed can also be used to factor expressions in the form $ax^2 + bxy + cy^2$. The only difference is that the final form of the answer is $(ex + fy)(gx + hy)$ instead of $(ex + f)(gx + h)$.

Example 4: Factor $12x^2 + 11xy + 2y^2$.

- Our goal is to factor this trinomial into two binomials $(ex + fy)(gx + hy)$.

- $(ex + fy)(gx + hy)$ expands into $egx^2 + (eh + fg)xy + fhy^2$. It follows that eg equals 12, $eh + fg = 11$, and $fh = 2$.

- We test whole-numbered values of the variables and try to find a combination that satisfies all of the requirements.

- Since the only pair of positive integers that multiply to two is 1 and 2, we set $f = 1$ and $h = 2$. If we switch the two values the obtained binomials will be exactly the same but in reverse order.

- When $e = 3$, $f = 2$, $g = 4$, and $h = 1$, $eh + fg = 11$. Therefore, this trinomial can be factored into $(3x + 2y)(4x + y)$.

Let us say that $(ex + f)(gx + h) = 0$ for some values of e, f, g, and h.

This equation has up to two solutions.

For two numbers to multiply to 0, one of the numbers has to be 0. No matter what value the other number holds, the product will still be 0.

Therefore, either $ex + f$ or $gx + h$ has to be 0 for the equation to hold true. The two solvable equations $ex + f = 0$ and $gx + h = 0$ determine the solutions to the original equation.

The above method can be used to solve some quadratic equations, which are equations of the form $ax^2 + bx + c = \ldots$ for real values a, b, and c. The highest degree of any term in a quadratic equation is 2.

Example 5: Solve $(2x + 5)(4x + 1) = 0$.

- The product of two values equals 0 if one or more of the values is 0. Therefore, this equation is satisfied when $2x + 5 = 0$ and when $4x + 1 = 0$.

- Solving these equations, we find that the solutions are $x = -\dfrac{5}{2}$ and $-\dfrac{1}{4}$.

Problems: Factoring

1 Bronze. Factor $(2x^4 + 4x^2 + 12x)$.

2 Bronze. Simplify $\dfrac{3x + 9}{x + 3}$.

3 Silver. Solve the equation $2x^2 + 6x = 0$.

4 Silver. Factor the expression $(x^2 + 6x + 9)$.

5 Silver. Solve the equation $x^2 + 6x + 9 = 0$.

6 Silver. Factor the expression $6x^2 + 20x + 16$.

7 Silver. Solve the equation $x^2 + 5x + 4 = 0$.

8 Silver. Bobelina is thinking of a number greater than -3. She then adds six times her number to 9, and then adds the square of the original number to her sum. Her result is 4. What was Bobelina's original number?

9 Silver. If $2x - 32 = 2y$, what is the value of $x - y$?

10 Silver. Simplify $\dfrac{3x^2 + 17x + 10}{x + 5}$.

11 Silver. I have four metal bars. Each bar has its length increased by 24%. By what percentage has the sum of the lengths of the four metal bars increased?

12 Gold. Factor $x^4 + 4x^2 + 4$.

13 Gold. Consider the 3 real numbers x, y, and z. If the sum of their squares is 15 and $xy + yz + xz = 10$, find the square of their sum.

14 Platinum. What is the sum of the roots of $x^2 + 987x - 153$? Note that the roots of an equation are the solutions when set to equal 0.

Part 18: Special Factorizations

There are a few special factorizations that should just be memorized.

(1) $a^2 - b^2 = (a + b)(a - b)$
(2) $a^3 + b^3 = (a + b)(a^2 - ab + b^2)$
(3) $a^3 - b^3 = (a - b)(a^2 + ab + b^2)$

The factorization $a^2 - b^2 = (a+b)(a - b)$ has many applications. One of them is simplifying radical expressions.

For a fractional term to be fully simplified, there should be no radicals in its denominator. The process of removing the radicals is known as rationalizing the denominator.

Consider $\dfrac{2}{\sqrt{3}}$. In order to rationalize the denominator, one would multiply the fraction by $\dfrac{\sqrt{3}}{\sqrt{3}}$. Doing so squares the radical in the denominator and simplifies the fraction to $\dfrac{2\sqrt{3}}{3}$ without changing its value. This is because $\dfrac{\sqrt{3}}{\sqrt{3}}$ equals 1, and multiplying something by 1 does nothing to its value.

Example 1: Simplify $\dfrac{5}{2\sqrt{11}}$.

- $\sqrt{11} \times \sqrt{11} = 11$. If we multiply the fraction by $\dfrac{\sqrt{11}}{\sqrt{11}}$, its value will not change, since $\dfrac{\sqrt{11}}{\sqrt{11}} = 1$.

- $\dfrac{5}{2\sqrt{11}} \times \dfrac{\sqrt{11}}{\sqrt{11}} = \dfrac{5 \times \sqrt{11}}{2 \times \sqrt{11} \times \sqrt{11}} = \dfrac{5\sqrt{11}}{22}$.

Example 2: Simplify $\dfrac{1}{\sqrt{3} + \sqrt{5}}$?

- By using the factorization of $a^2 - b^2$, we can square both $\sqrt{3}$ and $\sqrt{5}$ in a single step.
- The denominator is in the form $(a + b)$, so we must multiply by $(a - b)$ to make $a^2 - b^2$. Therefore, we must multiply

the denominator by $\sqrt{3} - \sqrt{5}$. We must also multiply the numerator by $\sqrt{3} - \sqrt{5}$ so that the value of the whole term does not change.

- Doing so, we obtain $\dfrac{\sqrt{3} - \sqrt{5}}{3 - 5}$ or $\dfrac{\sqrt{3} - \sqrt{5}}{-2}$, which can be simplified to $\dfrac{\sqrt{5} - \sqrt{3}}{2}$ by imposing the negative in -2 on the numerator.

Problems: Special Factorizations

1 Silver. Simplify $\dfrac{10}{\sqrt{10}}$ so that there are no radicals in the denominator.

2 Silver. Simplify $\dfrac{\sqrt{6}}{\sqrt{12}}$.

3 Silver. Find the value of $51^2 - 49^2$ without writing anything down.

4 Gold. Rationalize the denominator of $\dfrac{1}{\sqrt{6} + \sqrt{10}}$.

5 Gold. Simplify $\dfrac{\sqrt{10} + \sqrt{6}}{\sqrt{10} - \sqrt{6}}$.

6 Silver. If $(a + b) = 3$ and $ab = 4$, find $a^3 + b^3$.

7 Silver. If $(a - b) = 10$ and $ab = 6$, find $a^3 - b^3$.

8 Gold. Factor $a^2 + 2ab + b^2 - c^2$ into the product of two trinomials.

Part 19: The Quadratic Formula

Besides factoring, there is another way to solve quadratic equations. There exists a 1-size-fits-all formula known as the quadratic formula that allows you to solve for all quadratics, even those that can't be factored into nice numbers.

For a quadratic equation in the form $ax^2 + bx + c = 0$ for some values of a, b, and c, the solutions to the equation are $x = \dfrac{-b + \sqrt{b^2 - 4ac}}{2a}$ and $x = \dfrac{-b - \sqrt{b^2 - 4ac}}{2a}$.

It is extremely helpful to memorize this formula.

Example 1: Find all solutions of $x^2 + 4x + 4 = 0$ using the quadratic formula.

- In this quadratic expression, a is 1, b is 4, and c is also 4.
- Therefore, the solutions to this equation are
$$x = \frac{-4 + \sqrt{4^2 - 4(1)(4)}}{2(1)} \text{ and } x = \frac{-4 - \sqrt{4^2 - 4(1)(4)}}{2(1)}.$$
- These simplify to $x = \dfrac{-4 + \sqrt{0}}{2}$ and $x = \dfrac{-4 - \sqrt{0}}{2}$, which further simplify to -2 and -2. We see that there is only one solution to this equation: $x = -2$.

Quadratic equations can have 0, 1, or 2 solutions.

Note that you cannot take the square root of a negative number. Therefore, if $b^2 - 4ac$ is negative, the quadratic equation has no solution.

If $b^2 - 4ac$ is positive, the quadratic equation has two solutions.

If $b^2 - 4ac = 0$, the solutions to the quadratic equation $ax^2 + bx + c$ are $\dfrac{-b + 0}{2a}$ and $\dfrac{-b - 0}{2a}$. Both are equivalent to $\dfrac{-b}{2a}$, so the quadratic equation has one distinct solution.

How would we go about proving the quadratic formula? The proof of the quadratic formula involves a process called completing the square, in which we turn an expression of the form $x^2 + bx + c$ into an expression of the form $(x + d)^2 + e$ for some constants a, b, c, d, and e. Unfortunately, it also involves some pretty nasty algebra.

Example 2: The expression $x^2 + 4x + 5$ is equivalent to $(x + d)^2 + e$ for some values of d and e. Find d and e.

- Let us expand the expression $(x + d)^2 + e$. Doing so yields $x^2 + 2dx + d^2 + e$.
- For this to be equivalent to $x^2 + 4x + 5$ for all values of x, $2dx$ must correspond to $4x$ and $d^2 + e$ must correspond to 5. Therefore, $2d = 4$ and $d^2 + e = 5$.
- Solving the first equation, we find that $d = 2$. Substituting this value into the second equation yields $4 + e = 5$. Therefore, $e = 1$.

Let us first try to solve the quadratic $2x^2 + 7x + 3 = 0$ by completing the square.

- To set this equation up for completing the square, we first divide both sides by 2. Doing so yields $x^2 + \frac{7}{2}x + \frac{3}{2} = 0$.

- To make completing the square even easier, let us subtract $\frac{3}{2}$ from both sides. Doing so yields $x^2 + \frac{7}{2}x = -\frac{3}{2}$. This step is not essential, but it makes the process easier.

- We now set $x^2 + \frac{7}{2}x$ equivalent to $(x + d)^2 + e$, where d and e are constants.

- Expanding $(x + d)^2 + e$ yields $x^2 + 2dx + d^2 + e$. For this expression to be equivalent to $x^2 + \frac{7}{2}x$ regardless of the value that x takes up, $2dx$ must correspond to $\frac{7}{2}x$ and $d^2 + e$ must correspond to 0. Therefore, $2d = \frac{7}{2}$ and $d^2 + e = 0$.

- Dividing both sides of the first equation by 2 yields $d = \frac{7}{4}$. Plugging this value into $d^2 + e = 0$ yields $\frac{49}{16} + e = 0$, and therefore $e = -\frac{49}{16}$.

- We now have $\left(x + \frac{7}{4}\right)^2 - \frac{49}{16} = -\frac{3}{2}$. Adding $\frac{49}{16}$ to both sides yields $\left(x + \frac{7}{4}\right)^2 = \frac{25}{16}$.

- Let us look at this logically. This equation will be satisfied if $\left(x + \frac{7}{4}\right) = \sqrt{\frac{25}{16}}$ and if $\left(x + \frac{7}{4}\right) = -\sqrt{\frac{25}{16}}$ because in both cases $\left(x + \frac{7}{4}\right)^2$ will equal $\frac{25}{16}$.

- $\sqrt{\frac{25}{16}} = \frac{\sqrt{25}}{\sqrt{16}} = \frac{5}{4}$, so the two equations are now $x + \frac{7}{4} = \frac{5}{4}$ and $x + \frac{7}{4} = -\frac{5}{4}$. Therefore, the two solutions are $x = -\frac{1}{2}$ and $x = -3$.

Let us consider the quadratic equation $ax^2 + bx + c = 0$.

- To set this equation up for completing the square, we divide both sides by a. Doing so yields $x^2 + \frac{b}{a}x + \frac{c}{a} = 0$. To make completing the square even easier, let us subtract $\frac{c}{a}$ from both sides. Doing so yields $x^2 + \frac{b}{a}x = -\frac{c}{a}$.

- Our goal now is to turn $x^2 + \frac{b}{a}x$ into $(x + d)^2 + e$, where d and e cannot be numerical values but can be expressions in terms of a and b.

- Let us set $x^2 + \frac{b}{a}x$ and $(x + d)^2 + e$ as equivalent expressions. We also know that $(x + d)^2 + e$ is equivalent to $x^2 + 2dx + d^2 + e$.

- Since $x^2 + \frac{b}{a}x$ and $x^2 + 2dx + d^2 + e$ are equivalent for all values of x, $2dx$ must correspond to $\frac{b}{a}x$ and $d^2 + e$ must correspond to 0. It follows that $2d = \frac{b}{a}$ and $d^2 + e = 0$.

- $d = \frac{b}{2a}$, and we can plug this in to $d^2 + e = 0$ to find e. Doing so yields $\frac{b^2}{4a^2} + e = 0$, and therefore $e = -\frac{b^2}{4a^2}$.

- We now have that $\left(x + \frac{b}{2a}\right)^2 - \frac{b^2}{4a^2} = -\frac{c}{a}$. Adding $\frac{b^2}{4a^2}$ to both sides yields $\left(x + \frac{b}{2a}\right)^2 = \frac{b^2}{4a^2} - \frac{c}{a}$.

- Let us try to collapse the right side of the equation into one term using common denominators. How do we turn a into $4a^2$? We will multiply both the numerator and denominator of $\frac{c}{a}$ by $4a$, which is the same thing as multiplying the entire fraction by 1. Doing so yields $\frac{4ac}{4a^2}$.

- $\frac{b^2}{4a^2} - \frac{4ac}{4a^2}$ is the same as just $\frac{b^2 - 4ac}{4a^2}$, so the equation is now $\left(x + \frac{b}{2a}\right)^2 = \frac{b^2 - 4ac}{4a^2}$.

- Let us now take the square root of both sides. In actuality, doing so results in four equations, but two are identical. The equations are $\left(x + \dfrac{b}{2a}\right) = \sqrt{\dfrac{b^2 - 4ac}{4a^2}}$, $-\left(x + \dfrac{b}{2a}\right) = \sqrt{\dfrac{b^2 - 4ac}{4a^2}}$, $-\left(x + \dfrac{b}{2a}\right) = -\sqrt{\dfrac{b^2 - 4ac}{4a^2}}$, and $\left(x + \dfrac{b}{2a}\right) = -\sqrt{\dfrac{b^2 - 4ac}{4a^2}}$. However, only two of these equations are distinct. The two equations for x are $\left(x + \dfrac{b}{2a}\right) = \sqrt{\dfrac{b^2 - 4ac}{4a^2}}$ and $\left(x + \dfrac{b}{2a}\right) = -\sqrt{\dfrac{b^2 - 4ac}{4a^2}}$.

- $\sqrt{\dfrac{b^2 - 4ac}{4a^2}} = \dfrac{\sqrt{b^2 - 4ac}}{\sqrt{4a^2}} = \dfrac{\sqrt{b^2 - 4ac}}{2a}$. Our two equations are now $\left(x + \dfrac{b}{2a}\right) = \dfrac{\sqrt{b^2 - 4ac}}{2a}$ and $\left(x + \dfrac{b}{2a}\right) = \dfrac{-\sqrt{b^2 - 4ac}}{2a}$. Subtracting $\dfrac{b}{2a}$ from both sides in both equations yields the big formula: $x = \dfrac{-b + \sqrt{b^2 - 4ac}}{2a}$ and $= \dfrac{-b - \sqrt{b^2 - 4ac}}{2a}$.

Problems: The Quadratic Formula

1. Silver. Solve the equation $2a^2 + 9a + 7 = 0$.
2. Silver. Solve the equation $3a^2 - a - 12 = 0$.
3. Silver. Solve the equation $x^2 - 2x + 31 = 0$.
4. Silver. Find all x that satisfy $5x^2 + 10x - 3 = -8$.
5. Silver. Find all m that satisfy $3m^2 - m\sqrt{7} - 22 = 5$.
6. Silver. $4a^3 - 21a^2 - 10a + 23 = 5a^3 + 24 - 15a^2 + 21a - a^3$. Find all possible values of a.
7. Gold. $\dfrac{1}{c} + c = 4$. Solve for c.
8. Gold. $\dfrac{2x + 3}{x + 5} = \dfrac{3x + 2}{2x + 1}$. Solve for x.

9 Gold. Two positive real numbers have a sum of 17 and a product of 65. Find the two numbers.

10 Silver. The equation $2x^2 + bx + 10$ has exactly one real solution. Find all possible values of b.

11 Silver. How many real solutions does the equation $11x^2 + 22x + 15 = x + 4$ have?

Part 20: Solving for Whole Expressions

Many algebra problems ask us to find the values of expressions instead of single variables. This occurs frequently when we are given unsolvable equations or systems of equations where it is not possible to find the values of the individual variables. Solving these problems often takes considerable creativity.

Example 1: $3x + 4y + 3z = 10$ and $6x + 5y + 6z = 21$. Find the sum of x, y and z.

- We are trying to find $x + y + z$.
- If we add the two given equations, we obtain $9x + 9y + 9z = 31$.
- Factoring 9 out the left side yields $9(x + y + z) = 31$. Dividing both sides by 9, we find that $x + y + z = \dfrac{31}{9}$.

The hardest part of solving this problem is thinking to add the equations. Ideas like this will come to you with enough practice. Many of these problems involve trial and error, so keep trying different methods until something works.

Problems: Solving For Whole Expressions

1 Silver. $x = \dfrac{1}{2}y$ and $y = \dfrac{12}{13}z$. What fraction of z is x?

2 Silver. $3x + 4y = 12$ and $4x + 3y = 16$. Find $3x + 3y$. Try solving this problem without writing much down.

3 Gold. If $5x - 15y = 42$, what is the value of $-1\frac{1}{2}y + \frac{1}{2}x$?

4 Gold. $3x + 2y + 5z = 21$ and $x + 4y + 7z = 30$. Find the value of $x - y - z$.

5 Platinum. Solve the equation $x^3 + 3x^2 + 3x + 5 = 19$. Express your answer in simplest radical form.

6 Silver. $3x - y = 5x + 24y$. Find the value of $\frac{x}{y}$.

7 Gold. $xy + yz = 35$, $y + z = 21$ and $y = \frac{13}{x}$. Solve for y.

8 Gold. $a + b = 7$ and $b + c = 8$. Find the value of $a - c$.

Part 21: Infinite Series

What if you have a series of numbers that never ends? This series follows a specific pattern, and it goes on forever. Believe it or not, many infinite series can be simplified to a single value; we say that these series converge to the value. However, many infinite series also diverge, which means that they get infinitely large in the negative or positive direction.

Example 1: Simplify $1 + \frac{1}{2} + \frac{1}{4} + \frac{1}{8} + \frac{1}{16} \cdots$.

- This series follows the pattern that each term is one half of the previous term. Let us first set the value of this series equal to x. Now we have the equation $1 + \frac{1}{2} + \frac{1}{4} + \frac{1}{8} + \frac{1}{16} + \cdots = x$.

- Notice that if we multiply both sides of this equation by 2, we get the new series $2 + 1 + \frac{1}{2} + \frac{1}{4} + \frac{1}{8} + \frac{1}{16} + \cdots$, and this is equivalent to $2x$.

- What we can do now is substitute the first series into the second, since after the 2 in the second series, the rest is identical to the first series.

- Doing this, we receive $2 + x = 2x$. Therefore, $x = 2$.

- Let us now calculate the value of the series up to a certain point and see if the series appears to be converging to 2. $1 + \frac{1}{2} = 1.5$,

$$1 + \frac{1}{2} + \frac{1}{4} = 1.75, 1 + \frac{1}{2} + \frac{1}{4} + \frac{1}{8} = 1.875, 1 + \frac{1}{2} + \frac{1}{4} + \frac{1}{8}$$
$$+ \frac{1}{16} = 1.9375, \text{ and } 1 + \frac{1}{2} + \frac{1}{4} + \frac{1}{8} + \frac{1}{16} + \frac{1}{32} = 1.96875.$$

As you can see, this series gets closer and closer to 2 as we expand it further, but it never reaches 2. Most infinite series never actually reach the value that they converge to, but the difference between the sum of the partial series and the value the series converges to gets smaller and smaller forever.

The way to simplify most infinite series is to set the series equal to x and find a way to manipulate the series to get the same series again. This is the method we used in the first problem, and it can be used in other problems that look very different. Another thing to note is that in some series, there is no way to use this method.

In some series, the sum can keep increasing without approaching any specific value. The value of the series is infinity. Using clever algebra to simplify a series to a single value without making sure that the series does converge to a single value leads to very wrong answers.

Let us go back to finite geometric series. How can we prove that a geometric series with first term a, common ratio r, and n terms sums to $a \times \dfrac{1 - r^n}{1 - r}$?

- The sum of the first n terms of a geometric series is $a + ar + ar^2 + ar^3 \ldots + ar^{n-1}$. Let us set this equal to S, so that $S = a + ar + ar^2 + ar^3 \ldots + ar^{n-1}$.

- If this series was infinite, we would try to manipulate it to receive the same series again. This could be accomplished by multiplying both sides by r.

- Let us see what happens when we do this to a finite geometric series. Multiplying both sides of $S = a + ar + ar^2 + ar^3 \ldots + ar^{n-1}$ by r, we obtain $r \times S = ar + ar^2 + ar^3 \ldots + ar^{n-1} + ar^n$ Note that ar^n is equivalent to $a \times (r^n)$, not $(ar)^n$.

- There is a large amount of overlap between $ar + ar^2 + ar^3 \ldots + ar^{n-1} + ar^n$ and $a + ar + ar^2 + ar^3 \ldots + ar^{n-1}$. What happens when we subtract these two equations?

- Subtracting the two equations yields $S - (r \times S) = (a + ar + ar^2 + ar^3 \ldots + ar^{n-1}) - (ar + ar^2 + ar^3 \ldots + ar^{n-1} + ar^n)$, which can be simplified to $S - (r \times S) = a - ar^n$.

To solve for S, we factor the right side into $S(1 - r)$, and then divide both sides by $(1 - r)$ to find that $S = \dfrac{a - ar^n}{1 - r}$, which is equivalent to $a \times \dfrac{1 - r^n}{1 - r}$.

There is a shortcut to find the sum of infinite *geometric* series, meaning that there is a common ratio between all consecutive terms of the series. The formula for the sum of an infinite geometric series is $\dfrac{a}{1 - r}$, where a is the first term and r is the common ratio. However, this formula only works when $-1 < r < 1$, otherwise the real answer is infinity or negative infinity.

We just showed that a geometric series with first term a, common ratio r, and n terms sums to $a \times \dfrac{1 - r^n}{1 - r}$. If r is between -1 and 1, r^n approaches 0 as n approaches infinity. Therefore, $a \times \dfrac{1 - r^n}{1 - r}$ becomes $a \times \dfrac{1 - 0}{1 - r}$ when n is infinitely large.

Now let us discuss fractions. Consider the fraction $\dfrac{1}{3}$. If we convert this fraction to a decimal, we would end up with $0.33333\ldots$, where the threes go on forever.

Example 2: Convert $0.3333\ldots$ into a fraction.

- Let us call $0.3333\ldots n$. If we multiply n by 10, we receive $3.3333\ldots$ This is simply $3 + n$. Therefore, $10n = 3 + n$ and $n = \dfrac{1}{3}$.

The general rule is that if one digit is being repeated indefinitely in a decimal number, we divide that digit by 9 to convert the decimal into a fraction, if two digits are being repeated indefinitely in a decimal number, we divide the number formed with the two digits by 99 to convert the decimal into a fraction, and so forth. For example, $0.2222\ldots = \dfrac{2}{9}$, $0.148148148148\ldots = \dfrac{148}{999}$, and $0.142857142857142857142857\ldots = \dfrac{142857}{999999} = \dfrac{1}{7}$. The more general "9-trick" can be proven in the same manner as the previous example.

If we are given a decimal such as $0.2456565656\ldots$, where other digits come before the repeated digits, we must use an equation that multiplies the decimal by a power of 10 to receive a decimal that can be used with the "9-trick."

In the case of 0.2456565656, we set it equal to n and multiply both sides by 100, obtaining the equation $100n = 24.56565656\ldots$, which can be simplified to $100n = 24\dfrac{56}{99}$. Solving this equation, we find that the desired fraction is $\dfrac{608}{2475}$.

Problems: Infinite Series

1 Silver. Simplify $8 + 4 + 2 + 1 + \dfrac{1}{2} + \dfrac{1}{4} + \cdots$.

2 Bronze. Write $0.13131313\ldots$ as a fraction.

3 Silver. Find the value of $10 + 2 + \dfrac{2}{5} + \dfrac{2}{25} + \dfrac{2}{125} + \dfrac{2}{625} + \cdots$.

4 Platinum. Find the value of

$$\sqrt{110 - \sqrt{110 - \sqrt{110 - \sqrt{110 - \sqrt{110}}}}}\ldots.$$

5 Platinum. Find the value of $\cfrac{1}{5 + \cfrac{1}{5 + \cfrac{1}{5 + \cfrac{1}{5 + 1\ldots}}}}$.

6 Silver. Find the value of $1 + 2 + 3 + 4 + 5 + 6 + 7\ldots$.

7 Platinum. Find the value of the infinite series $\dfrac{1}{3} \times \left(\dfrac{1}{2} + \dfrac{1}{3} \times \left(\dfrac{1}{2} + \dfrac{1}{3} \times \left(\dfrac{1}{2} + \dfrac{1}{3} \times \left(\dfrac{1}{2} \ldots \right) \right) \right) \right)$.

8 Gold. Find the reciprocal of $0.12555555\ldots$. Express your answer as a common fraction if necessary.

Part 22: Sets

A set is a collection of things. Sets can be of toys, burgers, numbers, variables, etc. An element is simply an item in a set. When sets of numbers or variables are notated, the elements are listed and separated by commas. The whole set is enclosed by { }.

Consider the sets {1,2,3,4,5} and {3,4,5,6}. The intersection ($\cap$) of these sets is the set of all their shared elements. The union ($\cup$) of these sets is the set of all the distinct elements included in either one of the original sets. The union of {1,2,3,4,5} and {3,4,5,6} is {1,2,3,4,5,6}, and the intersection is {3,4,5}.

Sets can be easily visualized with Venn Diagrams.

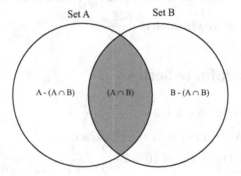

The gray region is the intersection of Set A and Set B. The white region on the left is Set A without the intersection, and the white region on the right is Set B without the intersection. Let us call the number of elements in Set A a, the number of elements in the intersection b, and the number of elements in set B c. It is easy to see that $(a - b) + (c - b) + b$ or $a + c - b$ equals the number of elements in the union of Set A and Set B.

One important fact to note is that for any two sets, the number of elements in the first set plus the number of elements in the second set minus the number of elements in the intersection equals the number of elements in the union. This can be seen with the Venn Diagram.

Example 1: In a building, there are 220 people. All of them either have a watch, a phone, or both. If 110 people have phones, and 20 have both a phone and a watch, how many people have only a watch?

- There are two different sets in this problem: the set of all the people with a watch and the set of all the people with a phone. The phone set has 110 elements, the intersection of the sets has 20 elements, and the union of the sets has 220 elements.

- Let us call the number of people with watches w, and the number of people with phones p. We already know that $p = 110$, so we need the value of w. From the set analysis we did before, we know that $110 + w - 20 = 220$. Therefore, $w = 130$.

- Here is a Venn Diagram that represents the scenario.

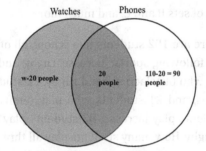

Watches   Phones

w-20 people   20 people   110-20 = 90 people

The three regions add up to 220 people

- From this we can easily see that $w - 20 + 20 + 90 = 220$, which yields $w = 130$.

We now know how to do problems using two sets, but what about three? There are seven distinct parts to a group of three sets: the intersection of all three sets, the three intersections of the pairs of two sets, and the three individual sets.

Here is a three-set Venn Diagram. You can easily see where the intersections are and how they relate to the rest of the sets.

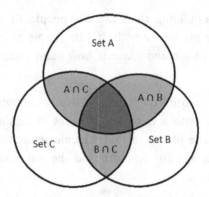

Let us define each region in the diagram as a specific variable. The white regions of sets A, B, and C will be a, b, and c, respectively. The light gray regions of A $\cap$ B, B $\cap$ C, and A $\cap$ C will be d, e, and f, and the dark gray region, the intersection of all three sets, will be g.

It is easy to see that $a + b + c + d + e + f + g$ equals the union of sets A, B and C. We can also find that $a + d + f + g =$ Set A, $e + g =$ the intersection of sets B and C, and much more.

Example 2: There are 152 students in a school, all of whom play at least one of the following sports: lacrosse, rugby, and tennis. There are 20 students who play lacrosse and rugby, 22 students who play rugby and tennis, and 21 students who play tennis and lacrosse. In total, 58 students play lacrosse, 82 students play tennis and 67 students play rugby. How many students play all three sports?

- There are three sets in this problem: the set of all the students that play lacrosse, the set of all the students that play rugby, and the set of all the students that play tennis. We'll call the intersection of all three sets x.

- There are 20 students who play lacrosse and rugby in total (including those who play tennis), so there are $20 - x$ students who play lacrosse and rugby but not tennis. Similarly, there are $22 - x$ students who play rugby and tennis but not lacrosse, and there are $21 - x$ students who play tennis and lacrosse but not rugby.

- To find the number of students who play just lacrosse, we have to take the 58 total students who play lacrosse and subtract away the students who play lacrosse and rugby, lacrosse and tennis, and all three. Doing this, we find that there are $58 - (20 - x) - (21 - x) - x = 17 + x$ students who play only lacrosse.

- Similarly, there are $82 - (22-x) - (21-x) - x = 39+x$ students who play only tennis, and $67 - (20-x) - (22-x) - x = 25+x$ students who play only rugby.

- We have now found all seven parts of the group of three sets. Now, we add all of these and set the sum equal to the 152 total students in the school. Doing this, we receive the equation $x + (20-x) + (21-x) + (22-x) + (17+x) + (39+x) + (25+x) = 152$. Solving, we find that $x = 8$ students.

- Here is a Venn Diagram that represents this scenario:

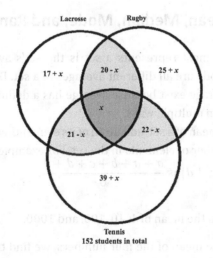

Tennis
152 students in total

Problems: Sets

1 Bronze. Find $\{a, c, d, f\} \cup \{a, c, g\}$.

2 Bronze. Find $\{a, c, d, f\} \cap \{a, c, g\}$.

3 Bronze. The intersection of the sets $\{a, b, c\}$ and $\{a, d, e\}$ is $\left\{\dfrac{1}{2}\right\}$. Find the value of a.

4 Silver. There are 200 students at Water Fountain High School. 120 students own tubas, 98 own saxophones, and 30 own neither. How many students at Water Fountain own both a tuba and a saxophone?

5 Gold. A certain food shop serves burgers, drinks, and appetizers. Last Tuesday, they received 120 customers, all of whom purchased at least one of the three items. 60 customers purchased drinks, 60 purchased burgers, and 49 purchased appetizers. 14 customers purchased burgers and appetizers, 21 purchased drinks and appetizers, and 20 purchased burgers and drinks. How many customers purchased all three?

Part 23: Mean, Median, Mode, and Range

The value that most represents a set is the set's average. Mean, median, and mode are all different averages of a set. Different ways to calculate average exist because average has a definition that can be interpreted in multiple ways.

To take the mean of a set, add up all of its elements and divide this sum by the number of elements in the set. For example, the mean of the set $\{a, a, b, c, d, d\}$ is $\dfrac{a + a + b + c + d + d}{6}$.

Example 1: Find the mean of 1, 10, 100, and 1000.

- To take the mean of the four numbers, we find their sum and divide it by 4:
  $(1 + 10 + 100 + 1000)/4 = 277.75$

To take the median of a set, list the elements in numerical order and find the middle element. For example, the median of the ordered set $\{a, b, c, d, e\}$ is c, because c is right in the middle.

In an ordered set with 7 elements, the median will be the fourth, in an element with 13 elements, the median will be the seventh, and so on.

If the set has an even number of elements, the set has no exact middle; instead, there are two numbers that the exact middle, if it existed, would be in between. To find the median of the whole set, take the mean of those two middle terms.

Example 2: Find the median of the set $\{1, 3, 2, 2, 9, 10\}$.

- The first step is to rearrange the set into numerical order: $\{1, 2, 2, 3, 9, 10\}$.

- The set has an even number of elements, so there is no term exactly in the middle. The two middle terms are 2 and 3.

- We take the mean of 2 and 3, which is $\dfrac{2+3}{2} = 2.5$

The mode of a set is the element that occurs most often.

For example, the mode of the set $\{a, a, b, c, c, d, c\}$ is c, since there are more occurrences of c than of any other element.

In the set $\{a, a, b, b, c\}$, there are two occurrences of both a and b, so the set has two modes: a and b.

If the *unique* mode of a set is given, it is the only mode.

Example 3: Find the mode of the set $\{1, 2 + 2, 8^2/4^3, 3 \times (4/3), 3 - 2\}$.

- Simplifying each expression, the set becomes $\{1, 4, 1, 4, 1\}$. 1 occurs three times and 4 occurs twice, so 1 is the mode of the set.

A set's range is the positive difference between its greatest and least terms.

Example 4: Find the range of the set $\{1, 2, 1, 21, 10\}$.

- The greatest element is 21 and the least element is 1. $21 - 1 = 20$.

Problems: Mean, Median, Mode, and Range

1 Bronze. Find the difference between the median and the mean of the set $\{a, b\}$.

2 Silver. The mean of the ordered set $\{a, b, c\}$ is equivalent to the sum of its range and its median. Find a in terms of b and c.

3 Silver. There are four jars on a shelf. Each jar contains some number of marbles. The mean number of marbles per jar among the four jars is 25. One more jar is going to be added to the shelf so that the mean number of marbles per jar becomes 30. How many marbles are in this jar?

4 Silver. Find the median, mode and range of the set $\{9, 5, 10, 12, 1, 2, 6, 5, 14\}$.

5 Gold. A set made up of 13 positive integers has a mean of 7 and a median of 8. What is the greatest possible positive integer that can be in this set?

6 Gold. Ino had an average score of 93% on her first three math tests. If she wants to bring her average test score up to at least a 97%, and it is possible for her to get a 100% (but no greater) on every forthcoming test, what is the minimum number of additional tests that Ino needs to take to do this?

7 Silver. Consider an increasing arithmetic sequence with n terms, common difference d, and first term a. Find the range of the sequence in terms of n and d.

Solutions Manual

Part 1: Linear Equations

1 Bronze.

- We subtract 4 from both sides to obtain x alone. Here we find that $x = 5$.

2 Bronze.

- Subtract 3 from both sides to get $3x$ alone.
- Divide both sides by 3. Here you find that $x = 10$.

3 Bronze.

- Subtracting 5 from both sides removes 5 from $a + 5$ and isolates the variable a. From here $a = 3$.

4 Bronze.

- First, we subtract 45 from both sides. Now we have $4x = 4$.
- Then, we divide both sides by 4 to find that $x = 1$.

5 Bronze.

- We want the coefficient of x to be 1. To do this, we have to divide both sides by $\frac{9}{19}$, since any real number divided by itself equals 1.
- Dividing by a fraction is the same thing as multiplying by the fraction's reciprocal. Therefore, we multiply both sides of the equation by $\frac{19}{9}$ to isolate x.
- Doing this, we find that $x = 38$.

6 Bronze.

- First, we subtract 12.1 from both sides. This leaves us with $2.5x = -7.5$.
- We divide both sides by 2.5 to find that $x = -3$.

7 Bronze.

- The first step in isolating the variable is collapsing all the x's into one term. To do this, we subtract x from both sides.
- We then obtain the equation $4 = 3 + x$. We subtract 3 from both sides to find that $x = 1$.

8 Bronze.

- To solve this equation, we first subtract 14.92 from both sides. This leaves $15.38x = 261.46$.

- Next, we divide both sides of the equation by 15.38. Doing so yields $x = 17$.

9 Bronze.

- First, we subtract $10c$ from both sides, obtaining $-\frac{7}{2}c + 4 = 9$.
- Subtracting 4 from both sides yields $-\frac{7}{2}c = 5$, and multiplying both sides by $-\frac{2}{7}$, we find that $c = -\frac{10}{7}$.

10 Bronze.

- Let us call the number of Joharu's coins j and the number of Bebi's coins b.
- Since Joharu and Bebi have 24 coins in total, $j + b = 24$.
- We also know that Joharu has 18 coins, so $j = 18$. It follows that $18 + b = 24$.
- Subtracting 18 from both sides of the equation yields $b = 6$, so Bebi has 6 coins.

11 Bronze.

- Let us call the required number of visitors p. $5p$ is the amount of money the amusement park makes from ticket sales. This amount must equal 500 dollars, so we set up the equation $5p = 500$.
- Dividing both sides by 5 yields $p = 100$, so 100 people must visit the amusement park for it to make 500 dollars.

12 Silver.

- Let us call Ashok's number of trading cards a. Since Shekar has 22 trading cards, their combined number of trading cards is $22 + a$.
- The problem gives us that Ashok has $\frac{1}{3}(22 + a) - 4$ cards, so $a = \frac{1}{3}(22 + a) - 4$.

- Distributing $\frac{1}{3}$ across the terms in parentheses yields $a = \frac{22}{3} + \frac{1}{3}a - 4$. Simplifying the right side by subtracting 4 from $\frac{22}{3}$ yields $a = \frac{1}{3}a + \frac{10}{3}$.

- Subtracting $\frac{1}{3}a$ from both sides yields $\frac{2}{3}a = \frac{10}{3}$.

- We then divide both sides by $\frac{2}{3}$ (a.k.a. multiply both sides by $\frac{3}{2}$) to find that $a = 5$ trading cards.

13 Bronze.

- $(4y + 3) - (2y + 1)$ is equivalent to $4y + 3 - 2y - 1$ due to the distributive property. This expression can be further simplified into $2y + 2$ by combining like terms.

- We now have the equation $2y + 2 = 42$.

- Subtracting 2 from both sides yields $2y = 40$, and dividing both sides by 2 yields $y = 20$.

14 Bronze.

- $(a + 3)/5$ can be written as $\frac{a}{5} + \frac{3}{5}$. Also, $\frac{a}{5} = \frac{1a}{5} = \frac{1}{5}a$.

- Implementing these two facts, we rewrite the given equation as $\frac{1}{5}a + \frac{3}{5} = 9$.

- Subtracting $\frac{3}{5}$ from both sides yields $\frac{1}{5}a = \frac{42}{5}$. Multiplying both sides by 5, $\frac{1}{5}$'s reciprocal, we find that $a = 42$.

15 Bronze.

- Let us call x the number of 150-mile segments that the person can travel. Each one of the segments costs \$30.

- The initial fee of \$100 plus the amount of money spent on the segments should equal the total amount of money that the person has, \$520. We can set up the equation $100 + 30x = 520$.

- Solving this equation, we find that $x = 14$ 150-mile segments. Therefore, the person can travel $14 \times 150 = 2100$ miles.

16 Bronze.

- Vinod obtains $2 \times 3 = 6$ smaller pieces of paper for each larger sheet that he purchases.

- If we call x the number of large sheets of paper that Vinod purchases, $6x$ is the number of small pieces that he receives.

- We set up the equation $6x = 84$ to solve for x.

- Dividing both sides of the equation by 6, we find that $x = 14$ sheets of paper.

17 Bronze.

- Let us call the number x.

- x doubled and added to 5 equals $5 + 2x$. Since this is one third of the original number x, we can set up the equation $5 + 2x = \frac{1}{3}x$.

- Subtracting $2x$ from both sides yields $-\frac{5}{3}x = 5$, and dividing both sides by $-\frac{5}{3}$ (or multiplying both sides by $-\frac{3}{5}$) we find that $x = -3$.

18 Bronze.

- Although this equation has two variables, subtracting $2z$ from both sides removes all z's from the equation, leaving just $x = 3$.

19 Bronze.

- Let us call the number x.

- Twelve subtracted from x can be written as $x - 12$. Since this is equivalent to twice the number's value, $x - 12 = 2x$.

- Subtracting x from both sides to isolate the variable, we find that $x = -12$.

20 Silver.

- If we distribute 2 over the parentheses on the right side, we obtain the equation $x + 12 = x + 3$.
- Subtracting 3 and then x from both sides, we obtain $0 = 9$.
- This is a universal untruth, so the equation has no solution.

21 Silver.

- $2(9-x)$ equals $18-2x$, and $4(4.5-0.5x)$ also equals $18-2x$.
- We have that $18 - 2x = 18 - 2x$. This is a universal truth, so all real numbers satisfy the equation.

22 Gold.

- Let us call n the year they met up.
- The older woman is the woman that was born in 1940. In the year n, the older woman's age is $n - 1940$. The younger woman's age is $n - 1957$.
- We have that $n - 1940 = 2(n - 1957) + 1$. Solving this equation, we find that $n = 1973$.

23 Bronze.

- To isolate x, we must eliminate the coefficient 2, the 4 being subtracted from $2x$, and the 7 dividing the whole left side. Which will we get rid of first?
- When solving equations, we must act on whole sides, not just individual terms. It is more difficult to remove the 2 and the 4 when the 7 is still on the left side.
- Therefore, we multiply both sides of the equation by 7 first. $7\left(\dfrac{2x-4}{7}\right) = 7(4)$ simplifies to $2x - 4 = 28$, since the 7s cancel on the left side. $2x - 4 = 28$ is a simple equation where $x = 16$.

24 Bronze.

- One way to find $2x + 9$ would be to solve for x in the given equation and then plug x in.

- Another way to find $2x+9$ would be to manipulate the current equation.
- In $x + 3 = 5$, the way to receive a $2x$ is to multiply both sides by 2. Doing so yields $2x + 6 = 10$.
- To get to $2x + 9$, we add 3 to both sides. This leaves us with $2x + 9 = 13$

Part 2: Cross Multiplication

1 Bronze.

- First, we multiply both sides of the equation by $2x$, obtaining $3 = \dfrac{4x}{8}$.
- Then, we multiply both sides of the equation by 8, obtaining $4x = 24$.
- Dividing both sides by 4, we find that $x = 6$.

2 Bronze.

- First, we multiply both sides of the equation by x. Doing so yields $4 = 10x$.
- Dividing both sides by 10, we find that $x = \dfrac{4}{10}$ or $\dfrac{2}{5}$.

3 Silver.

- To solve this equation, we must eliminate the denominators of both fractional expressions.
- To remove the denominator on the left side of the equation, we multiply both sides by $9 + x$, obtaining $9 = \dfrac{12(9 + x)}{19 + x}$.
- Next, we remove the other denominator by multiplying both sides by $19+x$, obtaining $9(19 + x) = 12(9 + x)$.
- Distributing on both sides yields the equation $171+9x = 108+ 12x$. Subtracting $9x$ and then 108 from both sides, we find that $3x = 63$, and therefore $x = 21$.

4 Silver.

- Let us start off by expanding both the numerator and denominator of the fraction on the left side of the equation. $4(3 + x) = 12 + 4x$, and $3(9 - x) = 27 - 3x$.

- We then multiply both sides by $27 - 3x$, obtaining $12 + 4x = \frac{8}{3}(27 - 3x)$, or $12 + 4x = 72 - 8x$.

- Solving this equation, we find that $x = 5$.

5 Silver.

- Let us call the amount of water that the worker put into the second pipe w. $w + 200$ is the amount of water that traveled to the filter.

- The filter halved this amount, so the number of gallons that came out of the filter is $\frac{w + 200}{2}$.

- The factory tripled this amount, so the amount of water that came out of the factory is $3\left(\frac{w + 200}{2}\right)$.

- This amount equals 939 gallon, so we can set up the equation $3\left(\frac{w + 200}{2}\right) = 939$.

- The first step in isolating w is dividing both sides by 3. This leaves us with $\frac{w + 200}{2} = 313$. Next, we multiply both sides by 2. Doing so yields $w + 200 = 626$, and therefore $w = 426$ gallon.

6 Bronze.

- To end up with an easily solvable equation, we must get the variable z out of the denominators of the fractions. One way to do this is to multiply both sides of the equation by z.

- $z\left(\frac{1}{2z} + \frac{5}{6z}\right) = 12(z)$ simplifies to $\frac{1}{2} + \frac{5}{6} = 12z$, which further simplifies to $\frac{4}{3} = 12z$. Dividing both sides by 12, we obtain our answer, $z = \frac{1}{9}$.

- Another way to solve this problem would be to use common denominators. If we try and change $\dfrac{1}{2z}$ so that its denominator is $6z$, we could combine it with $\dfrac{5}{6z}$ to produce one large term. This is the same principle that is used when adding and subtracting fractions.

- The way to make $\dfrac{1}{2z}$ have $6z$ as its denominator is to multiply both its numerator and denominator by 3. This won't change the value of the fraction, since we are essentially multiplying the fraction by 1, which doesn't change the fraction's value.

- Doing so yields $\dfrac{3}{6z} \cdot \dfrac{3}{6z} + \dfrac{5}{6z} = \dfrac{8}{6z}$. Now we have that $\dfrac{8}{6z} = 12$, an equation that is much easier to solve through cross multiplication. This equation also yields that $z = \dfrac{1}{9}$.

7 Silver.

- We first collapse the denominator of the given expression into one term.

- Using a common denominator of 30, $\dfrac{x}{3} = \dfrac{10x}{30}, \dfrac{y}{2} = \dfrac{15y}{30}$, and $\dfrac{1}{5} = \dfrac{6}{30}$ The sum of the three is $\dfrac{10x + 15y + 6}{30}$.

- $\dfrac{2x}{\dfrac{10x + 15y + 6}{30}} = 2x \times \dfrac{30}{10x + 15y + 6} = \dfrac{60x}{10x + 15y + 6}$.

Part 3: Systems of Equations

1 Silver.

- Let us use substitution to solve this problem. We will solve for y in terms of x first.

- Subtracting x from both sides in the first equation yields that $y = 7 - x$. Next, we replace y in the second equation with this expression.

- Now we have that $2x + (7 - x) = 11$. Simplifying this further, we find that $x + 7 = 11$, so $x = 4$. Plugging this value of x back into the first equation yields that $4 + y = 7$, so $y = 3$.

2 Bronze.

- Let us use elimination to solve this problem. This system is already set up to eliminate the variable y, so we add the two equations.

- $(x + y) + (x - y) = 50 + 10$ simplifies into $2x = 60$, so $x = 30$.

- Plugging this value of x back into the first equation, we obtain $30 + y = 50$, so $y = 20$.

3 Silver.

- Let us use substitution to solve this problem. This time, we will solve for x in terms of y. Subtracting $2y$ from both sides in the second equation, we find that $x = 16 - 2y$.

- Substituting this back into the first equation yields $2(16 - 2y) + y = 20$.

- Distributing the 2 across the parentheses, we expand the equation into $32 - 4y + y = 20$. Solving, we find that $y = 4$.

- Plugging this value back into the first equation, we find that $2x + 4 = 20$, so $x = 8$.

4 Silver.

- Let us use elimination to solve this system. To set up the variable a to be canceled, we have to multiply the entire second equation by -2.

- Doing so, we get our new second equation, $-3a - b = -5$.

- After adding this equation to the first equation, we find that $4b = 7$. Therefore, $b = \dfrac{7}{4}$.

- Plugging this value of b back into the first equation yields $3a + \dfrac{35}{4} = 12$. Solving, we find that $a = \dfrac{13}{12}$.

5 Silver.

- Let us call the amount of money a pencil costs in cents a and the amount a pen costs b. We have that $2a + 3b = 58$ and $4a + 4b = 100$.

- Let us use elimination to solve this system of equations. To set up the variable a to be eliminated, we multiply the first equation by -2, receiving $-4a - 6b = -116$.

- Next, we add the two equations, obtaining $(4a + 4b) + (-4a - 6b) = 100 - 116$. This simplifies into $-2b = -16$, so $b = 8$.

- Substituting this value back into the first equation yields that $2a + 24 = 58$, so $a = 17$. Therefore, a pencil costs 17 cents.

6 Silver.

- Let us call the amount of money a bottle of water costs in dollars w, and the amount a bottle of syrup costs s.

- We have that $16w + 37s = 90$, and that $8w + 23s = 54$. Again, let us use elimination to solve this system.

- To set up the variable w to be eliminated, we multiply the second equation by -2, obtaining $-16w - 46s = -108$.

- Now we add the two equations, which gives us $-9s = -18$, and therefore $s = 2$. Plugging this value back into the first equation, we find that $w = 1$. Therefore, a bottle of water costs 1 dollar.

7 Gold.

- In this problem, our strategy will be to find both a and b in terms of c, and then substitute these expressions into the first equation. Let us start with the third equation, where we will find a in terms of c.

- Subtracting c from both sides, we find that $a = 11 - c$.

- Now we move on to the second equation. Here we also subtract c from both sides, obtaining $b = 9 - c$.

- Now we substitute both of these expressions into our first equation, obtaining $(11 - c) + (9 - c) = 6$. Simplifying, we find that $20 - 2c = 6$, and therefore $c = 7$.

8 Bronze.

- $xy + xy = 2xy$. xy and xy are identical terms, so they can be added just like single variables. We now have that $2xy = 32$, so $xy = 16$.

9 Silver.

- Let us call the number of camels with one hump A, the number of camels with two humps B, and the number of camels with three humps C.

- The problem already tells us that $B = 10$. Knowing this, we have that $A + 10 + C = 21$, so $A + C = 11$.

- We also know that there are 45 humps in total, so $(1)A + (2)10 + (3)C = 45$.

- From this we find that $A + 3C = 25$. We now have a system of equations that can be solved using substitution.

- Solving for A in terms of C in the first equation yields that $A = 11 - C$. Plugging this value into the second equation, we find that $11 - C + 3C = 25$. Solving this equation, we find that $C = 7$.

- Plugging this value back into $A + C = 11$, we find that $A = 4$. Therefore, there are 4 camels with one hump in the zoo.

10 Gold.

- Since y is 10 when x is 4, we can substitute these values of x and y into $y = mx + b$ to form the equation $10 = 4m + b$.

- We also have that y is 19 when x is 7, so we can substitute these values into $y = mx + b$ as well to form the new equation $19 = 7m + b$.

- These two equations form a solvable system. We can derive that $b = 10 - 4m$ from $10 = 4m + b$, and then we can substitute this expression for b in the next equation to make $19 = 7m + (10 - 4m)$.

- Simplifying this equation, we find $19 = 3m + 10$, and therefore $m = 3$.

- Substituting m with 3 in the equation $10 = 4m + b$, we obtain $10 = 12 + b$, and therefore $b = -2$.

11 Silver.

- Let us use substitution to solve this system.

- Multiplying both sides by y and then by x in the first equation yields that $30x = 25y$.

- Dividing both sides of this by 30 to solve for x in terms of y, we find that $x = \dfrac{5}{6}y$.

- Next, we substitute $\dfrac{5}{6}y$ for x in the second equation, $x + 2y = 85$, which leaves us with $\dfrac{5}{6}y + 2y = 85$.

- Solving, we find that $y = 30$. Substituting this value of y back into $x + 2y = 85$ yields $x + 2(30) = 85$, and therefore $x = 25$.

12 Gold.

- Our goal in this problem is to isolate the variable c. The first step is to subtract ab from both sides.

- This leaves us with $cd = ef - ab$.

- To get rid of the d multiplying the c, we must divide both sides by d. This gives us our answer: $c = \dfrac{ef - ab}{d}$.

13 Gold.

- If we distribute the c across the parentheses, we can rewrite $c(c + d)$ as $c^2 + cd$.

- To isolate the variable d, we have to first subtract c^2 from both sides. This leaves us with $cd = 13 - c^2$.

- Next, we have to divide both sides by c to get rid of the c multiplying the d. Doing so yields $d = \dfrac{13 - c^2}{c}$, our answer.

14 Bronze.

- $abcd + 4abcd = 5abcd$, as $abcd$ and $abcd$ are identical terms. We now have that $5abcd = 10$, so $abcd = 2$.

Part 4: Roots and Exponents

1 Bronze

- $9^2 = 9 \times 9 = 81$.

2 Silver

- $-3^4 = -(3 \times 3 \times 3 \times 3) = -81$, and $3/2^2 = \dfrac{3}{4}$. Therefore, the answer is $-81 + \dfrac{3}{4}$ or $-\dfrac{321}{4}$.

3 Bronze

- The square roots of 64 are 8 and -8. Simple square roots like this should just be memorized.

4 Bronze

- By our exponent laws, $140^{23} \times 140^x = 140^{23+x}$.
- Since we have that this equals 140^{35}, $23 + x$ must be the same thing as 35. Therefore, $x = 12$.

5 Bronze

- The cube root of 8 is 2.

6 Bronze

- -1 raised to any odd-numbered power always turns out to be -1. Therefore, $a = -1$.

7 7 Bronze.

- Since $a^5 = -32$, $a = \sqrt[5]{-32}$. Fifth roots have only one solution, since five is odd.
- Just like square roots, the fifth root of an integer is either an integer as well or irrational. Hopefully, the fifth root of -32 turns out to be an integer.

- We also know that a is negative, since a positive number raised to the fifth power cannot be negative.

  Testing the fifth powers of negative integers in order starting from $(-1)^5 = -1$, we find that $(-2)^5 = -32$, so $\sqrt[5]{-32} = -2$. Therefore, $a = -2$.

8 Bronze

- $(3xy)^2$ can be expanded into $3^2 \times x^2 \times y^2$, which can be simplified into $9x^2y^2$.

9 Bronze

- $12 \times 12 = 144$, so the answer is 12. (Radical signs only require positive solutions.)

10 Bronze

- To simplify this problem, we first take the square root of both sides. This leaves us with $a^2 = \pm 9$. -9 has no square root, and 9's only positive square root is 3. Therefore, $a = 3$.

11 Bronze

- Dividing both sides by x, we get that $x^2 = 4$. Therefore, $x = 2$, since we are just asked for the positive value.

12 Bronze

- To isolate the variable x, we must divide both sides by x^{99}. Once we do this, we are left with $7 = x$, our answer.

13 Silver

- The solutions to this equation are the positive and negative square roots, $+\sqrt{2585214}$ and $-\sqrt{2585214}$. The sum of these is 0, since the sum of any number and its negative is 0.

14 Bronze

- The positive solution to $x^2 = 169$ is 13, since $13 \times 13 = 169$. This is another fact that should just be memorized.

15 Silver

- This problem can be also thought of as the square root of the square root of 1296. Taking the square root of 1296 with a calculator yields 36, and taking the square root of 36 yields 6 as our answer. (We don't have to include the negative value because this problem has a radical sign.)

16 Bronze

- First, we split this problem up into $(x^2/x^3)(y^2/y^5)$. $x^2/x^3 = x^{2-3} = x^{-1} = 1/x$. $y^2/y^5 = y^{2-5} = y^{-3} = 1/y^3$.

- Multiplying the two, we obtain our answer, $1/xy^3$.

17 Bronze

- First, we split this problem up into $(a^3/a^{-3})(b^2/b^4)(c^4/c^{-2})$. $a^3/a^{-3} = a^{3-(-3)} = a^6$, $b^2/b^4 = b^{2-4} = b^{-2} = 1/b^2$, and $c^4/c^{-2} = c^{4-(-2)} = c^6$.

- Multiplying the two, we obtain $a^6 c^6/b^2$.

18 Bronze

- This product can be broken up into $\frac{3}{2} \times 9^8/9^5$. $9^8/9^5 = 9^{8-5} = 9^3 = 729$, and $729 \times \frac{3}{2} = \frac{2187}{2}$.

19 Silver

- The square root of 50 is greater than 7 but less than 8, since 7^2 is 49, 8^2 is 64, and 50 is in between 49 and 64.

- Similarly, the square root of 150 is greater than 12 but less than 13, since 12^2 is 144, 13^2 is 169, and 150 is in between 144 and 169.

- The whole numbers in between these two values are therefore 8 through 12, which makes 5 whole numbers.

20 Silver

- Vinod starts out with two pieces of paper. When he cuts each of these into three pieces, he obtains 2×3 total pieces. If he cuts each of these into three more pieces, he ends up with $2 \times 3 \times 3$ in total.

- If Vinod repeats this process five times, he will end up with $2 \times 3 \times 3 \times 3 \times 3 \times 3$ or 2×3^5 total pieces of paper. $3^5 = 243$, and $2 \times 243 = 486$ pieces.

21 Bronze

- Since we have a negative exponent, we first change the base, 9, into its reciprocal, $\frac{1}{9}$.

- Then, we raise $\frac{1}{9}$ to the one half power (a.k.a. take its square root). $\left(\frac{1}{3}\right)^2$ and $\left(-\frac{1}{3}\right)^2 = \frac{1}{9}$, so our answers are $\frac{1}{3}$ and $-\frac{1}{3}$.

22 Bronze

- We can combine $16^{3/4} \times 16^{1/2}$ into $16^{(3/4+1/2)}$ or $16^{5/4}$. $16^{5/4}$ can be split up into $(16^{1/4})^5$.

- $16^{1/4} = 2$ and -2, $2^5 = 32$, and $(-2)^5 = -32$. Therefore, our answers are 32 and -32.

23 Bronze

- What we must do is replace n with 4 and m with 5 in the formula that defines @. This leaves us with $4^2/(5^2 + 4)$, or $\frac{16}{29}$.

24 Bronze

- According to the order of operations, we first have to simplify what is in the parentheses first. Doing so yields $-(3)^4$.

- Exponents comes before multiplication, so we first have to simplify 3^4 before adding the negative on. $3^4 = 81$, so the answer is -81.

25 Bronze

- If -2 is raised to an odd-numbered power, there will be an odd number of negatives being multiplied which will result in a negative number.

- If -2 is raised to an even-numbered power, there will be an even number of negatives being multiplied which will result in an overall positive number.

- Since 155 is odd, $(-2)^{155}$ is negative.

26 Bronze

- Just as $25^{1/3} \times 5^{1/3}$ can be combined into $(25 \times 5)^{1/3}$ (using the fourth combining exponents law taught in this section),

$^3\sqrt{25} \times^3 \sqrt{5}$ can be combined into $\sqrt[3]{(25 \times 5)}$ or $\sqrt[3]{125}$, which equals 5.

- All roots can be combined like this, as long as they are the same degree root.

27 Bronze

- $\sqrt{\dfrac{625}{9}}$ can be broken up into $\sqrt{625}/\sqrt{9}$. $\sqrt{625}$ is 25 ($25 \times 25 = 625$), and $\sqrt{9}$ is 3.

- Therefore, the answer is $\dfrac{25}{3}$.

28 Bronze

- Our goal is still to isolate the variable n. Subtracting 8 from both sides yields $2n^2 = 8$, and dividing both sides by 2 yields $n^2 = 4$.

- If we take the square root of both sides, we obtain the two solutions $n = 2$ and $n = -2$.

29 Silver

- The way to obtain an a^5 in our equation is to raise both sides to the $\dfrac{1}{5}$ power, as $(a^{25})^{\frac{1}{5}} = a^{25 \times \frac{1}{5}} = a^5$. However, our calculations will be much easier if we divide both sides by $\dfrac{2}{3}$ a.k.a. multiply both sides by $\dfrac{3}{2}$ first.

- Doing so, we obtain $a^{25} = 243$. $243 = 3^5$, so $243^{\frac{1}{5}}$ is 3. Therefore, $a^5 = 3$.

30 Silver

- $64 = 8^2$. Therefore, $64^{2x} = (8^2)^{2x}$. By our properties of exponents, this equals $8^{2 \times 2x}$ or 8^{4x}. We now have that $8^{4x} = 8^{11}$, so $4x$ must equal 11. If $4x = 11$, $x = \dfrac{11}{4}$.

31 Silver

- Let us try to isolate $(x + 3)^2$ before isolating x, as we do not know what to do with this expression at this point.

- First, we multiply both sides of the equation by 6. This yields $(x+3)^2 + 19 = 50$. Subtracting 19 from both sides yields $(x+3)^2 = 31$.

- The first step in isolating x from here is taking the square root of both sides. Doing so yields the two equations $x + 3 = \sqrt{31}$ and $x + 3 = -\sqrt{31}$, since the square root of $(x+3)^2$ is $x + 3$ and the square root of 31 is $\pm\sqrt{31}$. Including the negative square root of $(x+3)^2$ will not give us any unique solutions, as the resulting negative sign on the left side can be canceled by multiplying both sides by -1. Therefore, we can ignore the negative square root of $(x+3)^2$ for now.

- Solving these two equations yields the two solutions $x = \sqrt{31} - 3$ and $x = -\sqrt{31} - 3$.

32 Silver

- Again, let us isolate $(x-3)^5$ before isolating x.

- The first step is to multiply both sides by 5. Doing so yields $(x-3)^5 + 3 = 35$. Subtracting 3 from both sides leaves $(x-3)^5 = 32$.

- Now, to isolate x, we must take the fifth root of both sides, a.k.a. raise both sides to the one-fifth power. The fifth root of $(x-3)^5$ is $(x-3)$, and the fifth root of 32 is 2. Therefore, $x - 3 = 2$, and $x = 5$.

Part 5: Simplifying Radical Expressions

1 Silver.

- $\sqrt{x^4} = x^2$, so the first thing we can do is take the x^4 out of the radical and instead write x^2 as a coefficient. Doing so yields $x^2\sqrt{y^3}$.

- $y^3 = y^2 \times y$, so $\sqrt{y^3} = \sqrt{y^2} \times \sqrt{y} = y\sqrt{y}$.

- After going through these steps, our final answer is $x^2 y\sqrt{y}$.

2 Bronze.

- 4 is a perfect square, and $84/4 = 21$. $\sqrt{84} = \sqrt{4} \times \sqrt{21} = 2\sqrt{21}$.
- 21 has no perfect square factors except for 1, so we are done.

3 Silver.

- $288 = 144 \times 2$. $\sqrt{144} = 12$, so our answer is $12\sqrt{2}$

4 Silver.

- If we go through all of the perfect squares that are small enough to be factors of 182, we find that none of them divide 182 evenly. $\sqrt{182}$ is already in simplest form.

5 Gold.

- 1, 8, and 27 are the three smallest perfect cubes.
- $54 = 27 \times 2$, so $\sqrt[3]{54}$ can be split up into $\sqrt[3]{27} \times \sqrt[3]{2}$, which equals $3\sqrt[3]{2}$ (Since $3^3 = 27$).

6 Gold.

- 1, 16, and 81 are the three smallest perfect fourth powers.
- $810 = 81 \times 10$, so $\sqrt[4]{810}$ can be broken up into $\sqrt[4]{81} \times \sqrt[4]{10}$. $\sqrt[4]{81} = 3$, so our answer is $3\sqrt[4]{10}$.

7 Silver.

- 30^2 is easily recognizable as 900. This is pretty close to 950.
- $31^2 = 961$. 950 is between 900 and 961, and it is much closer to 961 than it is to 900. Therefore, 961 is the closest perfect square to 950.

Part 6: Proportions

1 Bronze.

- If x is inversely proportional to y, $x \times y$ has a constant value.
- Therefore, $3x$ multiplied by what y becomes must equal $x \times y$.
- In order for this to happen, y must become $\frac{1}{3}y$.

2 Bronze.

- The ratio of birds to eggs in the first scenario is $\dfrac{5}{10}$, or $\dfrac{1}{2}$. This ratio will remain constant even if the number of birds or the number of eggs changes.

- We set up the equation $\dfrac{10}{x} = \dfrac{1}{2}$, where x is what the problem asks for. To solve this equation, we first multiply both sides by x. This leaves us with $\dfrac{1}{2}x = 10$, and therefore $x = 20$ eggs.

3 Bronze.

- Here, the ratio of birds to eggs is $\dfrac{16}{24}$ or $\dfrac{2}{3}$.

- We set up the equation $\dfrac{2}{3} = \dfrac{20}{x}$, where x is what the problem asks us to find. Solving this equation, we find that $x = 30$ eggs.

4 Silver.

- If x is inversely proportional to y and directly proportional to z, $\dfrac{xy}{z}$ has a constant value.

- Substituting in the values of $x, y,$ and z that we are given in the problem, we find that this constant value equals $1(4)/5$ or $\dfrac{4}{5}$.

- Now we set up the equation $10(x)/12 = \dfrac{4}{5}$, which will yield the value of x in the new scenario. Solving, we find that $x = \dfrac{24}{25}$.

5 Gold.

- Through logic and reasoning, we determine that as the number of workers increases, the number of jobs increases and the number of days it takes to complete the jobs decreases.

- Therefore, the number of workers is directly proportional to the number of jobs and inversely proportional to the number of days. The constant value of $\dfrac{\text{workers} \times \text{days}}{\text{jobs}} = \dfrac{15}{2}$, which we derive from the first scenario given to us.

- In the new scenario, there are 4 workers and 6 jobs, so we set up the equation $\dfrac{4 \times d}{6} = \dfrac{15}{2}$, where d is the number of days it takes to complete the six jobs.

- Solving this equation, we find that $d = \dfrac{45}{4}$ days.

6 Silver.

- The square of the length of the alien's hair times the alien's height has a constant value. This value equals $9^2(9)$ or 729. Calling the length of the hair of the alien that is four inches tall x, we set up the equation $4x^2 = 729$.

- Dividing both sides by 4 yields $x^2 = \dfrac{729}{4}$, and therefore $x = \sqrt{\dfrac{729}{4}}$. Take into account that length cannot be negative. $\sqrt{729}$ is 27 and $\sqrt{4}$ is 2, so x equals $\dfrac{27}{2}$ or 13.5.

7 Silver.

- If the ratio of x to y is 2:5, $\dfrac{x}{y} = \dfrac{2}{5}$. Multiplying both sides of this equation by y, we find that $x = \dfrac{2}{5}y$.

- Substituting this expression for x into the equation $2x + 3y = 57$ and solving, we find that $y = 15$.

8 Bronze.

- If a is directly proportional to b, $\dfrac{a}{b}$ has a constant value. This value equals 3/1 or 3. When b is 98, $a/98$ still equals 3. Therefore, $a = 294$.

9 Silver.

- If a is directly proportional to b and inversely proportion to c^2, ac^2/b has a constant value. This value equals $21(2^2)/12$ or 7, found by substituting in the values we are given in the first scenario of the problem.

- When $b = 28$ and $c = 7$, $(7^2)a/28 = 7$. Therefore, $a = 4$.

10 Bronze.

- If a person weighs 100 pounds on Earth and 38 on Mars, the ratio of a person's weight on Mars to their weight on Earth is therefore $\dfrac{38}{100}$ or $\dfrac{19}{50}$.
- To find the weight of the person who weighs 100 pounds or Mars, we set up the equation $\dfrac{100}{x} = \dfrac{19}{50}$, where x is his or her weight on Earth. Solving this equation, we find that x to the nearest whole number equals 263 pounds.

11 Silver.

- The number of plants divided by the cube of the amount of oxygen in the room has a constant value. This value equals $8/2^3$ or 1.

- We set up the equation $27/x^3 = 1$, where x is the amount of oxygen in the room with 27 plants. Solving, we find that $x = 3$ pounds of oxygen.

12 Silver.

- The fraction of the total box which is red pencils is $\dfrac{7}{7+11}$ or $\dfrac{7}{18}$. Therefore, the fraction of the total box which is blue pencils is $\dfrac{11}{18}$, because the fraction of red pencils plus the fraction of blue pencils has to equal 1, the whole.
- $\dfrac{11}{18}$ of 90 equals $\dfrac{11}{18} \times 90$ or 55, so there are 55 blue pencils in the box.

13 Gold.

- Let us call the total number of items in the first bag x and the total number of items in the second bag y.
- The number of pencils in the first bag is $\left(\dfrac{9}{9+2}\right) x$ or $\dfrac{9}{11}x$, and the number of erasers in the first bag is $\dfrac{2}{11}x$. The number

of pencils in the second bag is $\frac{6}{11}y$, and the number of erasers in the second bag is $\frac{5}{11}y$.

- It follows that $\frac{9}{11}x + \frac{6}{11}y = 57$, and $\frac{2}{11}x + \frac{5}{11}y = 31$. This is a solvable system of equations.

- Let us simplify both of these equations by multiplying both sides by 11 in each one. Doing so, we obtain the new system $9x + 6y = 627$ and $2x + 5y = 341$.

- The problem asks us to find the value of y, and solving this system yields that $y = 55$ total items.

14 Silver.

- On Vinoe, the amount of land in square miles is $\frac{3}{3+5} \times 752$ or 282, and the amount of water is $\frac{5}{3+5} \times 752$ or 470.

- The amount of land in square miles on Eoniv is $\frac{2}{2+9} \times 374$ or 68, and the amount of water is $\frac{9}{2+9} \times 374$ or 306.

- The total amount of land on both planet and moon in square miles is $282 + 68$ or 350, and the total amount of water is $470 + 306$ or 776.

- Therefore, the combined ratio of water to land is $\frac{776}{350}$ or $\frac{388}{175}$.

15 Silver.

- Calling the length of the unknown road x, we set up the proportional equation $\frac{45 \text{ feet}}{9 \text{ inches}} = \frac{x \text{ feet}}{12 \text{ inches}}$. (We specify units because the units are not constant throughout the problem.)

- $\frac{45 \text{ feet}}{9 \text{ inches}}$ can be simplified to $\frac{15 \text{ feet}}{3 \text{ inches}}$, and multiplying both the numerator and denominator by 4 gives us the proportion $\frac{60 \text{ feet}}{12 \text{ inches}}$. Therefore, $x = 60$ feet.

- This problem serves as an example to how problems with varying units can be solved. Treat the conflicting units like different variables when manipulating them.

16 Silver.

- The ratio of c to d can be written as $\dfrac{c}{d}$. Similarly, $4c{:}5d$ can be written as $\dfrac{4c}{5d}$. Notice that $\dfrac{4c}{5d}$ contains $\dfrac{c}{d}$, as $\dfrac{4c}{5d} = \dfrac{4}{5}\left(\dfrac{c}{d}\right)$.

- We replace $\dfrac{c}{d}$ with $\dfrac{9}{23}$ in this expression, obtaining $\dfrac{4}{5}\left(\dfrac{9}{23}\right)$ or $\dfrac{36}{115}$.

Part 7: Inequalities

1 Bronze.

- Subtracting a from both sides, we obtain $35 < a - 12$. Adding 12 to both sides, we obtain $47 < a$.

- This inequality read the other way is $a > 47$.

2 Bronze.

- Subtracting 34 from both sides, we obtain $-3x \geq 24$.

- Dividing both sides by -3, we obtain $x \geq -8$, but since we divided both sides of the inequality by a negative number, we must flip the sign of the inequality.

- This gives us our answer, $x \leq -8$.

3 Bronze.

- Subtracting 10 from both sides of the equation, we find that $x > 3$.

4 Silver.

- The first step here is to multiply both sides of the inequality by -4.

- Since we multiplied both sides of an inequality by a negative number, we must flip the sign of the inequality. This leaves us with $2x - 3 \geq -164$.
- Adding 3 to both sides, we obtain $2x \geq -161$, and dividing both sides by 2, we obtain $x \geq -\dfrac{161}{2}$.

5 Bronze.

- x must be greater than 3 and less than or equal to 12. The integers that satisfy this are 4, 5, 6, 7, 8, 9, 10, 11, and 12. This makes 9 integers.

6 Silver.

- Let us call this minimum age x. $\dfrac{x}{3} + 30$ is the number of aliens that he or she spawns. This has to be greater than or equal to 40 aliens, so we can set up the inequality $\dfrac{x}{3} + 30 \geq 40$.
- Subtracting 10 from both sides, we obtain $\dfrac{x}{3} \geq 10$, and multiplying both sides by 3, we find that $x \geq 30$.
- The minimum age that is greater than or equal to 30 years old is 30 years old, our answer.

7 Silver.

- Subtracting 3 from both sides of $x + 3 > 45$, we find that $x > 42$.
- Subtracting 10 from both sides of $2x + 10 \geq 40$ and then dividing both sides by 2, we find that $x \geq 15$.
- These are the two pieces of information that we have about x. If x is greater than 42, it automatically is greater than or equal to 15, so we can simplify this to just $x > 42$.

8 Silver.

- Subtracting 31 and then dividing by 2 on both sides of $2x + 31 > 32$, we obtain $x > \dfrac{1}{2}$.
- Adding 45 to both sides of $-3x - 45 > 5$, we obtain $-3x > 50$, and dividing both sides by -3 and flipping the sign of the inequality, we obtain $x < -\dfrac{50}{3}$.

- x cannot be greater than $\dfrac{1}{2}$ and less than $-\dfrac{50}{3}$ at the same time. Therefore, this system of inequalities has no solution.

9 Gold.

- Subtracting 24 from both sides of $12x + 24 \leq 156$, we obtain $12x \leq 132$, and dividing both sides by 12, we find that $x \leq 11$.

- Dividing both sides of $-45x < 157.5$ by -45 and then flipping the sign because we divided by a negative number, we obtain $x > -\dfrac{7}{2}$.

- There is a set portion of all numbers that satisfies both of these conditions for x, so we can write our answer as $-\dfrac{7}{2} < x \leq 11$.

10 Platinum.

- From this expression, we can derive three facts. $35 + x < 43 - 2x$, $43 - 2x \leq 5 - x$, and, if we ignore the $43 - 2x$ in the middle, we can also figure out that $35 + x < 5 - x$.

- Solving the first inequality we derived, we obtain $x < \dfrac{8}{3}$. Solving the second, we obtain $x \geq 38$. Solving the last, we obtain $x < -15$.

- If x is less than -15, it is automatically less than $\dfrac{8}{3}$, so we can ignore the $x < \dfrac{8}{3}$. We now have that $x \geq 38$ and that $x < -15$.

- Therefore, there is no solution to this inequality, because x cannot be greater than 38 and less than -15.

11 Gold.

- We cannot simply take the square root of both sides, obtaining $x > \pm 3$. What will make x^2 greater than 9?

- The first case is if $x > \sqrt{9}$, or $x > 3$. The second case is if x is less than the negative square root of 9, as for values such as $x = -4$, x^2 will be greater than 9.

- Therefore, our other solution is $x < -\sqrt{9}$, or $x < -3$. This is not similar to the systems of inequalities we have seen because it involves an or, not an and. Since x cannot be greater than 3 and less than -3 at the same time, our final answer is $x > 3$ or $x < -3$.

12 Platinum.
- To isolate x, we first have to multiply both sides of this inequality by b. However, since it is given that b is negative, we have to flip the sign of the inequality after doing so.

- This yields $x^2 < b(1 - a)$. We are also given that a is greater than 5, so $1-a$ will undoubtedly be negative. Since b is negative as well, $b(1 - a)$ will be positive.

- When will x^2 be less than this value? To do this, let us first think about when it will be greater. This will- happen when x is greater than the positive square root of $b(1-a)$, or $\sqrt{b(1-a)}$, or when x is less than the negative square root of $b(1 - a)$, or $-\sqrt{b(1-a)}$.

- For x^2 to be less than $b(1 - a)$, x has to be in between the two values that we found. Therefore, our solution is $-\sqrt{b(1-a)} < x < \sqrt{b(1-a)}$.

13 Gold.
- In an attempt to make it easier to isolate for x, let us split $\sqrt{2x}$ into $\sqrt{2} \times \sqrt{x}$. Now we can divide both sides by $\sqrt{2}$, obtaining the inequality $\sqrt{x} \le \dfrac{3}{\sqrt{2}}$.

- x cannot be negative, otherwise $\sqrt{x}$ is undefined. Therefore, there is only one case to this problem: x is less than or equal to $\left(\dfrac{3}{\sqrt{2}}\right)^2$. This yields the solution $x \le \dfrac{9}{2}$. To be more complete, we could have also included that x must be greater than or equal to 0.

14 Gold.
- Let us first compare $\dfrac{4}{57}$ and $\dfrac{2}{29}$. We write that $\dfrac{4}{57}(?)\dfrac{2}{29}$, where the (?) can be $<, =$ or $>$.

- Cross multiplying yields 4×29 (?) 57×2. Calculating both of these products, we find that the (?) is a $>$, so $\dfrac{4}{57}$ is greater than $\dfrac{2}{29}$.

- Next, let us compare $\dfrac{4}{57}$ and $\dfrac{3}{43}$ in the same way. We write that $\dfrac{4}{57}(?)\dfrac{3}{43}$, where the (?) can be $<, =$ or$>$.
- Cross multiplying yields 4×43 (?) 57×3. Calculating both of these products, we find that the (?) is a$>$, so $\dfrac{4}{57}$ is greater than $\dfrac{3}{43}$.

- Lastly, let us compare $\dfrac{2}{29}$ and $\dfrac{3}{43}$. We write that $\dfrac{2}{29}(?)\dfrac{3}{43}$, where the (?) can be $<, =,$ or$>$.
- Cross multiplying yields 43×2 (?) 29×3. Calculating both of these products, we find that the (?) is a $<$, so $\dfrac{2}{29}$ is less than $\dfrac{3}{43}$.

- $\dfrac{4}{57}$ is the greatest of the three fractions, as we have found that it is greater than the other two. Since $\dfrac{2}{29}$ is less than $\dfrac{3}{43}, \dfrac{2}{29}$ is the least of the three.
- Therefore, the answer is $\dfrac{2}{29}, \dfrac{3}{43}, \dfrac{4}{57}$.

Part 8: Counting Numbers

1 Bronze.
- The integers between 65 and 902 are all of the integers starting with 66 and ending with 901. This is a total of $901 - 66 + 1 = 836$ integers.

2 Bronze.
- There are $33321 - 445 + 1 = 32877$ integers in this set, since we include the 445.

3 Silver.
- The integers greater than 1 but less than or equal to 339 are the integers starting with 2 and ending with 339.
- This is a total of $339 - 2 + 1 = 338$ integers.

4 Silver.

- The first multiple of 8 strictly between 0 and 99 is 8, and the last is 96. These can also be written as 8(1) and 8(12).

- The other multiples of 8 that work are 8(2), 8(3), 8(4), etc. Therefore, the number of multiples of 8 strictly between 0 and 99 is the number of numbers from $1 - 12$, or 12 multiples.

5 Gold.

- It is very important to know certain facts about our months and years. Normal years have 365 days.

- However, every year that is divisible by 4 ($\dots 2000$, 2004, 2008, $\dots$) is a leap year, meaning that the year has 366 days. There are 12 months in a year. Every month except for February, April, June, September, and November has 31 days. April, June, September, and November have 30 days, and February has 28 days except for leap years, in which it has 29.

- From the beginning of December 3, 1989, to the end of December 31, 1989, $31 - 3 + 1$ or 29 days elapse. From here, we have to count the number of days from January 1 to June 9 of 1990.

- January has 31 days, February of 1989 has 28, March has 31, April has 30, and May also has 31. Therefore, from the beginning of January 1, the start of January, to the end of May 31, the end of May, $31 + 28 + 31 + 30 + 31 = 151$ days elapse.

- After May 31, there are 8 more days to go until the end of June 8th/the beginning of June 9. The total number of days that elapse is $29 + 151 + 8 = 188$ days.

Part 9: Sequences and Series

1 Bronze.

- Think of the sum of the first 15 positive integers as an arithmetic series with first term 1, last term 15, and common difference 1.

- The formula for the sum of an arithmetic series is $\dfrac{a+f}{2} \times n$, where a is the first term, f is the last term, and n is the number of terms. In this particular series, $a = 1$, $f = 15$, and $n = 15$.
- Plugging these values into the formula, we find that the sum of this series is 120.

2 Bronze.

- In this arithmetic series, $a = 1$, $f = 32$, and $n = 32$.
- Plugging these values into the formula for arithmetic series, we find that this sum is 528.

3 Silver.

- It is easy to see that $a = 3$ and $f = 47$ in the arithmetic series formula $\dfrac{a+f}{2} \times n$.
- But what is n, the number of terms? To find the position that 47 holds in the series, we have to use the arithmetic sequence formula: $s + (n - 1)d$. s, the starting term, is 3, and d, the common difference, is 4.
- We set up the equation $3 + 4(n-1) = 47$. Solving, we find that $n = 12$.
- 47 is the 12th term of the arithmetic series, and, subsequently, the series has 12 terms. Therefore, n in the arithmetic series formula is 12. Plugging these values into the formula, we find that the sum is 300.

4 Bronze.

- All arithmetic sequences can be modeled by $a, a + r, a + 2r, a + 3r, \ldots$, where a is the first term of the sequence and r is the common difference.
- Therefore, the n-th term is the first term added to $n - 1$ r's, or $a + r(n - 1)$. The problem gives us that a is 11, r is 4, and n is 16. Therefore, the term we are looking for is $11 + 4(15)$ or 71.

5 Bronze.

- This sequence is an arithmetic sequence with first term -7 and common difference 4. All arithmetic sequences can be modeled by $a, a + r, a + 2r, a + 3r, \ldots$, where a is the first term of the sequence and r is the common difference.

- Therefore, the n-th term is the first term added to $n - 1$ r's, or $a + r(n - 1)$. In this sequence, $a = -7, r = 4$, and $n = 25$, so the term we are looking for is $-7 + 4(24)$ or 89.

6 Bronze.

- The first 22 positive odd integers start at 1 and end at 43. Therefore, in our arithmetic series formula $\dfrac{a + f}{2} \times n$, $a = 1, f = 43$, and $n = 22$.

- Plugging these values in, we find that this sum is 484.

7 Bronze.

- The formula for the n-th term of an arithmetic sequence is $s + (n - 1)d$, where s is the starting term and d is the common difference. In this sequence, we have that $s = 3$ and $d = 6$.

- Therefore, the 10th term is $3 + (10 - 1)(6)$ or 57.

8 Gold.

- If the third term of an arithmetic sequence is 29, $s + 2d = 29$, where s is the first term and d is the common difference.

- If the eighth term is 38, $s + 7d = 38$.

- This leaves us with a solvable system of equations. Solving, we find that $d = 1.8$ and $s = 25.4$.

- The 15th term of the arithmetic sequence is $s + 14d$, or $25.4 + 14(1.8)$. This turns out to be 50.6.

9 Bronze.

- If the first term of a geometric progression is 2 and the second term is 3, the common quotient is $\dfrac{3}{2}$, because the first term multiplied by one common quotient has to equal the second term.

- The formula for the n-th term of a geometric sequence is $a \times q^{(n-1)}$, where a is the first term in the sequence and q is the common quotient. Therefore, the fifth term is $2 \times \left(\dfrac{3}{2}\right)^4$, or $\dfrac{81}{8}$.

10 Bronze.

- This sequence is a geometric sequence with first term 3 and common quotient 2. The seventh term of this sequence equals $3 \times (2)^{(7-1)}$, or 192.

11 Silver.

- Let us call the smallest of the five consecutive odd integers x. The next consecutive odd integer after x will be $x+2$, since consecutive odd integers are 2 apart.

- The integer after $x+2$ will be $x+4$, and then $x+6$, and lastly $x+8$. The sum of these integers is $x+(x+2)+(x+4)+(x+6)+(x+8)$, or $5x+20$ when simplified.

- We can now set up the equation $5x+20 = 115$, and therefore $x = 19$. The greatest two integers out of our five are $x+6$ and $x+8$, and now that we know $x = 19$, we find that these are 25 and 27 respectively. $25 \times 27 = 675$, our answer.

12 Silver.

- Let us number the teams 1 through 20. Team 1 plays teams 2–20 once each, which makes 19 games.

- We have already counted the game between Team 2 and Team 1, so we need to count Team 2 playing Teams 3–20, a total of 18 games.

- Similarly, Team 3 needs to play teams 4–20, which is 17 games. This pattern continues up to Team 19, which only has to play Team 20. The total number of games played is $1 + 2 + 3 + 4 + \cdots + 18 + 19$, or $\left(\dfrac{1+19}{2}\right)19$. This equals 190 games.

13 Bronze.

- In this series, the common difference is -4. The problem gives us that the first term is 1, the last term is -183, and the number of terms is 47.
- $\dfrac{1 + (-183)}{2} \times 47 = -4277$, our answer.

14 Gold.

- This is neither an arithmetic nor a geometric sequence. Notice that the numerators of the fractions start at 15 and continually increase by 5, and the denominators of the fractions start at 2 and continually increase by 1.

- Let us find an expression that determines the numerator of the n-th term, then an expression that determines the denominator, and put them together.

- The expression that determines the numerator is not $5n$, since at $n = 1$, the numerator already starts out at 15. Instead, it is $5(n + 2)$.

- The expression that determines the denominator is simply $n + 1$.

- Therefore, the answer is $\dfrac{5(n + 2)}{n + 1}$.

15 Silver.

- The key to this problem is to notice that the coefficients of x in the numerators of the fractions form an arithmetic sequence that is equivalent to the positive integers in order, and the denominators form an arithmetic sequence that is equivalent to the odd positive integers in order.

- It is easy to see to see that the numerator of the 502nd term is $502x$. The 502nd term of the sequence 1, 3, 5, 7, ... is $1 + 2(501)$ or 1003, as the first term is 1 and the common difference is 2.

- Therefore, the 502nd term of the original sequence is $\dfrac{502x}{1003}$.

16 Gold.

- It may be a little difficult to notice, but the difference between the first two terms is 2, the difference between the second and third is 3, the difference between the third and fourth is 4, and this pattern continues.

- Therefore, the second term is $1 + 2$, the third term is $1 + 2 + 3$, the fourth is $1 + 2 + 3 + 4$, etc.

- The 100th term of this sequence is therefore $1 + 2 + 3 + 4 + 5 + 6 + 7 + 8 + \cdots + 99 + 100$. This sum equals $\dfrac{1 + 100}{2} \times 100 = 5050$.

Part 10: Distance, Rate, and Time

1 Bronze.

- The cyclist's rate is 30 miles per hour, and the cyclist travels for three hours.

- Therefore, the distance that he travels is 30×3 or 90 miles.

2 Bronze.

- The distance this car travels is 50 miles, and the time it takes to travel this distance is 2 hours.

- We set up the equation $50 = 2r$, where r is the car's average speed (a.k.a. rate). Solving this equation, we find that $r = 25$ miles per hour.

3 Silver.

- This car's rate is 12.2 miles per hour, and the distance it is going to travel is 42.3 miles.

- We set up the equation $42.3 = 12.2t$, where t is the time that the car takes to travel the 42.3 miles.

- Solving this equation, we find that $t = 3.5$ hours to the nearest tenth.

4 Silver.

- Let us call the time that the faster runner runs for before catching up with the slower runner t.
- Since the slower runner gets a six minute head start, the slower runner runs for a total of $t + 6$ minutes.
- The faster runner travels $50t$ yards, and the slower runners travels $30(t + 6)$ yards.
- We set up the equation $50t = 30(t + 6)$, since both runners will have run the same total distance when they catch up with each other.
- Solving this equation, we find that $t = 9$ minutes.

5 Silver.

- This problem has a tricky aspect to it. We do not have to worry about the octagonal shaped building at all. All we need to know is that he runs for 30 minutes at 25 meters per minute.
- Therefore, his distance is 30×25 or 750 meters.

6 Platinum.

- We are given two different scenarios in this problem. The first scenario is the mess in the first three sentences, and the second scenario is given in the fourth sentence.
- Let us start with the second scenario. We'll call the distance from city A to city B d, and the time it takes for the train to get there t. Since *Distance = Rate × Time*, $d = 40t$.
- Now let us go back to the first scenario. The train travels $40 \times 1\frac{1}{2}$ or 60 miles before it breaks down. After it stops, it therefore has $d - 60$ miles left to travel.
- The train is exactly on time in both scenarios, so the total time in the scenarios are equivalent. It has already been on the road for $1\frac{1}{2}$ hours plus the 45 minutes or $\frac{3}{4}$ hours it was stopped, so it has $t - \left(1\frac{1}{2} + \frac{3}{4} \right) = t - \frac{9}{4}$ hours left to travel.

- In this time, it travels $50\left(t - \dfrac{9}{4}\right)$ miles, since it is traveling at 50 miles per hour.

- We now combine the distances traveled before and after the breakdown and write the equation $60 + 50\left(t - \dfrac{9}{4}\right) = d$. This equation is not in the format *Distance = Rate $\times$ Time*, it is in the format *Distance + More Distance = Total Distance*.

- We also have that $d = 40t$, so we can replace d with $40t$ in the equation. This leaves us with $60 + 50\left(t - \dfrac{9}{4}\right) = 40t$. Distributing the 50 across the parentheses yields $60 + 50t - \dfrac{225}{2} = 40t$.

- Solving this equation, we find that $t = \dfrac{21}{4}$ hours. Finally, we plug this value back into $d = 40t$ to obtain our answer, 210 miles.

7 Gold.

- These runners are traveling in opposite directions around a circular track, so when they meet up, the runners must have run one track length, or 100 feet, in total.

- The faster runner runs three times the distance of the slower runner in a set time, since he has a rate that is three times higher (try backing this statement with algebra).

- Let us call the slower runner's distance x. The faster runner's distance is therefore $3x$. We can set up the equation $x + 3x = 100$. Solving, we find that the slower runner runs 25 ft.

8 Silver.

- Since the second runner is two feet per minute faster than the first runner, the second runner gains two feet on the first runner every minute.

- To lap the first runner, the second runner needs to gain a total of 100 feet, or one track length, on the first.

- We can set up the equation $2t = 100$, where t is the amount of time it takes for this to happen. Solving, we find that it takes 50 minutes for the second runner to lap the first.

9 Gold.

- The answer is not simply $\dfrac{5 + 10}{2} = 7.5$ feet per second. This is because the runner travels at 5 feet per second for a longer time than he travels at 10 feet per second, so the two speeds do not hold equal weight in the average.

- It is essential to know that average speed equals *total* distance over *total* time. This jogger takes $50/5 = 10$ seconds to get to the second recreation center, and $50/10 = 5$ seconds to get back.

- Therefore, his total time is $10 + 5 = 15$ seconds. He runs 100 feet in total, since he travels 50 feet to the second recreation center and back. Therefore, his average speed is $100/15 = \dfrac{20}{3}$ feet per second.

- Remember to not just take the average of the rates when calculating average rate; go through this whole process.

Part 11: Rates

1 Silver.

- If a hose fills one bottle in three minutes, it fills $\dfrac{1}{3}$ of a bottle every minute. Therefore, its rate is $\dfrac{1}{3}$ in bottles per minute.

- Similarly, the rate of the second hose is $\dfrac{1}{2}$.

- Working together, the rate of the hoses is $\dfrac{1}{3} + \dfrac{1}{2}$ or $\dfrac{5}{6}$.

- Let us call the time that the hoses take to fill the bottle t. We set up the equation $\dfrac{5}{6}t = 1$, since the hoses are just going to be filling one bottle.

- Solving this equation, we find that it takes the hoses 1.2 minutes to fill the bottle.

2 Silver.

- The first machine's rate is $\frac{1}{32}$ toys per hour, and the second machine's rate is $\frac{1}{48}$ toys per hour. Working together, their rate is $\frac{1}{32} + \frac{1}{48} = \frac{5}{96}$ toys per hour.

- We set up the equation $\frac{5}{96}t = 1$, where t is the time it takes the machines to build one toy.

- Solving this equation, we find that this task takes the machines 19.2 hours.

3 Gold.

- The man's rate is $\frac{1}{10}$ roads per hour, and the combined rate of the man and his son is $\frac{1}{3}$ roads per hour.

- Let us call the son's rate r. We set up the equation $\frac{1}{10} + r = \frac{1}{3}$ to find r, and obtain $r = \frac{7}{30}$. If the son shovels $\frac{7}{30}$ roads in an hour, we can set up the equation $\frac{7}{30}t = 1$ to find the amount of time it takes for him to shovel one whole road alone. Solving, we find that this takes him $\frac{30}{7}$ hours.

4 Gold.

- Let us call Brock's rate of eating in pizzas per hour b and Sei's rate s. We have that $2b + s = 1$ and $4s + b = 1$.

- Solving this system of equations for b, we find that $b = \frac{3}{7}$.

- Since Brock eats $\frac{3}{7}$ of a pizza in an hour, he takes $1_{(\text{pizza})} / \frac{3}{7}$ or $\frac{7}{3}$ hours to eat a full pizza.

5 Silver.

- Each faucet has a rate of 0.6. Let us call the number of faucets that Vode is going to add x. Each time a faucet is added, the rate at which the tub fills up is going to increase by 0.6.

- Knowing this, we set up the equation $2(0.6x) = 30$. Solving this equation, we find that $x = 25$ faucets.

6 Gold.

- Hamie's rate in seats per minute is 4, and Erok's rate is 3.

- In the five minutes that Hamie cleans seats alone, he cleans $4(5) = 20$ seats. After Erok joins him, there are 35 seats left to clean.

- We set up the equation $t(3 + 4) = 35$, where t is the time it takes Hamie and Erok to finish cleaning the remaining seats and $(3 + 4)$ is their combined rate.

- Solving this equation, we find that $t = 5$ minutes.

- Including the five minutes in which Hamie cleans without Erok, Hamie cleans for $5 + 5 = 10$ minutes.

7 Silver.

- Both of these motors are simultaneously working to rotate the wheel. They counteract each other, since one is going clockwise and the other is going counter-clockwise.

- Their net rate is $(54 - 32)$ or 22 revolutions per second in the counter-clockwise direction.

- In 10 seconds, the wheel will rotate 22×10 or 220 times counter-clockwise.

8 Silver.

- The forces of the two teams counteract each other.

- If one team is pulling the rope east at 22 meters per second, and the other is pulling west at 20 meters per second, the 20 meters per second speed will cancel the 22 meters per second speed, but not completely. There will still be a $22 - 20 = 2$ meters per second net speed going east.

- It will take $10/2 = 5$ seconds for the rope to move 10 meters.

9 Gold.

- The minute hand of a clock makes one revolution per hour. The hour hand of a clock makes one revolution every 12 hours, so it makes $\frac{1}{12}$ of a revolution per hour.

- We set up the equation $x(1+\frac{1}{12}) = 1$, where x is the time it takes for the hands to sum to a full revolution. Solving this equation, we find that $x = \frac{12}{13}$ hours or $\frac{12}{13} \times 60 = \frac{720}{13}$ minutes.

- This amount of time past 4:00 equals 4:55 p.m. to the nearest minute.

10 Gold.

- After half an hour in the flashlight, the battery used up $\frac{1}{6}$ of its charge, since half an hour a.k.a. 30 minutes over 3 hours is $\frac{1}{6}$.

- Similarly, the battery used up $2\frac{1}{2}/6 = \frac{5}{12}$ of its charge in the fan. In total, it has now used up $\frac{1}{6} + \frac{5}{12} = \frac{7}{12}$ of its charge, so it has $\frac{5}{12}$ left.

- On a full or $\frac{12}{12}$ charge, the battery will last 12 hours in the speaker, so with $\frac{5}{12}$ of its charge it will last $\frac{5}{12}$ of 12 or 5 hours.

11 Platinum.

- The minute and hour hands are in the same position at 12:00 a.m. We visualize the time that they are going to be together again as sometime after 1:00 a.m., where the minute hand has made more than a full revolution and the hour and minute hands are both a little bit past the 1. This time should be slightly over 1:05.

- Let us define the rates of the minute and hour hands. The minute hand moves $\frac{1}{60}$ of a revolution per minute, and the hour hand moves $\dfrac{1}{60 \times 12}$ or $\dfrac{1}{720}$ of a revolution per minute, so these shall be their respective rates in revolutions per minute.

- The amount of time it takes for one hand to make n revolutions is n divided by the rate of the hand, according to the main formula of this section.

- In the scenario we are trying to analyze, the hour hand makes part of a revolution, and the minute hand makes a full revolution plus that same part.

- We write this in equation form as $(1+x)/\dfrac{1}{60} = x/\dfrac{1}{720}$, where x is the fraction of a revolution that we are trying to find. This equation can be simplified to $60(x + 1) = 720x$. Solving, we find that $x = \dfrac{1}{11}$ revolutions.

- To find the answer from here, we can either find what time corresponds to $\dfrac{1}{11}$ of a revolution of the hour hand past 12:00, or, since we already know that the hour is 1, we can find what $\dfrac{1}{11}$ of a revolution means for the minute hand and add that to 1:00 a.m.

- $\dfrac{1}{11}$ of 1 full revolution of the minute hand, or $\dfrac{1}{11}$ of 60 minutes, is 5 minutes and 27 seconds to the nearest second. Therefore, the answer is 1:05 a.m. and 27 seconds.

12 Gold.

- The first person cleans $\dfrac{1}{6}$ of a seat per minute, so his rate is $\dfrac{1}{6}$. The second person cleans 6 seats per minute, so that it his rate. (Remember that rate is the amount done in 1 unit of time.)

- Their combined rate is $\dfrac{1}{6} + 6$ or $\dfrac{37}{6}$ seats per minute, so we set up the equation $\dfrac{37}{6}t = 74$, where t is the amount of time it takes them to clean all of the seats in the auditorium. Solving, we find that $t = 12$ minutes.

13 Gold.

- It is easy to see that the minute hand is $\frac{1}{4}$ of a revolution into the clock at 4:15.

- The hour hand makes 1 full revolution every 12 hours. 4 hours and 15 minutes is $4\frac{1}{4}/12$ or $\frac{17}{48}$ of a full revolution of the hour hand, so the hour hand is $\frac{17}{48}$ of a revolution into the clock.

- The hour hand's rate is $\frac{1}{720}$ of a revolution per minute, and the minute hand's rate is $\frac{1}{60}$ of a revolution per minute. For the minute hand to catch up to the hour hand, it has to gain $\frac{17}{48} - \frac{1}{4} = \frac{5}{48}$ of a revolution on the hour hand.

- Every minute, the minute hand gains $\frac{1}{60} - \frac{1}{720} = \frac{11}{720}$ of a revolution on the hour hand. We set up the equation $\frac{11}{720}x = \frac{5}{48}$ to find how long this will take, and solving, we find that this time is $\frac{75}{11}$ or $6\frac{9}{11}$ minutes.

- $\frac{9}{11}$ of a minute equals $\frac{9}{11} \times 60$ or 49 seconds to the nearest second, so the answer is 6 minutes and 49 seconds past 4:15 or 4:21 and 49 seconds.

14 Silver.

- Let us use the unit $\frac{\text{increase of } y}{\text{increase of } x}$ for the rate of y's increase. y's rate with this unit is 3, as y increases by 3 per increase of 1 by x. If y increases by 42, let us call a the increase that x makes in return.

- We set up the equation $3a = 42$, as again, y increases by 3 per increase of 1 by x. Solving, we find that $a = 14$.

Part 12: Unit Conversion and Analysis

1 Silver.

- Our first step will be to convert 13 yams into roqs. To do this, we must set up the proportion $\dfrac{13 \text{ yams}}{x \text{ roqs}} = \dfrac{425 \text{ yams}}{1 \text{ roq}}$.

- Since our units match up, we can now disregard units and solve for x. Doing so, we find that $x = \dfrac{13}{425}$ roqs.

- We now convert this amount to gums. To do this, we set up the proportion $\dfrac{\frac{13}{425} \text{ roqs}}{n \text{ gums}} = \dfrac{1 \text{ roq}}{2125 \text{ gums}}$. Our units match up again, so we can disregard them and solve for n. Doing so, we find that $n = 65$ gums.

2 Silver.

- Let us convert the length of the current racetrack into yards. We already have that one mile is equivalent to 1760 yards.

- To convert 2.6 feet into yards, we can set up the proportion $\dfrac{2.4 \text{ feet}}{x \text{ yards}} = \dfrac{3 \text{ feet}}{1 \text{ yards}}$ (ignore the grammar issue).

- Since our units match up, we can disregard them and solve for x. Solving yields that $x = 0.8$ yards. We now have that this racetrack is $1760 + 2 + 0.8 = 1762.8$ yards long.

3 Bronze.

- The unit for mileage is miles per gallon, which can be written as $\dfrac{\text{miles}}{\text{gallon}}$. If we multiply by a quantity in gallons, the unit gallon should cancel and leave a quantity in miles.

- We set up the equation $25 \dfrac{\text{miles}}{\text{gallon}} \times x \text{ gallon} = 200$ miles. Solving algebraically, we find that $x = 8$ gallon

4 Gold.

- The desired answer is in the unit Joules. We have a quantity in grams, a quantity in degrees, and a value in $\dfrac{\text{Joules}}{\text{grams} \times^\circ \text{Celsius}}$.

- Treating units as variables, we find that we must multiply these three so that grams and degrees Celsius cancel, leaving just Joules.

- $10\,\text{grams} \times 10\,\text{degrees Celsius} \times \dfrac{8\,\text{Joules}}{\text{grams} \times °\,\text{Celsius}} = 800$ Joules, our answer.

5 Silver.

- We will first convert the car's speed to feet per hour, and then we will convert to miles per hour.

- There are 60 seconds in a minute, so there are 60×60 or 3600 seconds in 60 minutes. 60 minutes is equivalent to one hour, so there are 3600 seconds in an hour.

- In 3600 seconds, the car travels 20×3600 or 72,000 feet. It follows that the car's speed in feet per hour is 72,000.

- Since there are 5,280 feet in a mile, 72,000 feet is equivalent to $\dfrac{72000}{5280}$ or $\dfrac{150}{11}$ miles. The car's speed is therefore $\dfrac{150}{11}$ miles per hour.

Part 13: Percentages

1 Bronze.

- Since something percent means that something over 100, we can set up the equation $\dfrac{x}{100} = \dfrac{4}{5}$, where x is the percentage that we are trying to find.

- Solving this equation, we find that $x = 80$, so $\dfrac{4}{5} = 80\%$.

2 Bronze.

- Here we can set up the equation $\dfrac{x}{100} = \dfrac{3}{8}$ to find our desired percentage.

- Solving this equation, we find that it is 37.5%.

3 Bronze.

- 45% can be written as $\dfrac{45}{100}$. Simplifying, we get $\dfrac{9}{20}$.

4 Bronze.

- 50% can also be written as $\dfrac{50}{100}$, or $\dfrac{1}{2}$. $\dfrac{1}{2}$ of 320 is 160.

5 Silver.

- The amount of increase in bacteria is 25, and the original amount of bacteria is 50. Therefore, the percent increase is $(25/50) \times 100$, or 50%.

6 Gold.

- Each child gets $\dfrac{o}{c}$ cookies, since the o cookies are divided evenly among the c children.

- Out of the o cookies we started with, this represents $\dfrac{o}{c} \,/\, o$ of the total number of cookies.

- This can be simplified to $\dfrac{o}{c} \times \dfrac{1}{o}$, and further simplified to $\dfrac{1}{c}$.

- To make this into a percent, we have to multiply it by 100. Therefore, our answer is $\dfrac{100}{c}$%.

7 Gold.

- Let us call the original number of ants in the tree a. An increase by 30% to a can be written as $1.3a$.

- Three increases of 30% to a can be written as $(1.3)(1.3)(1.3)(a)$. This equals $2.197a$ when simplified.

- Now, we must find the percent increase from a to $2.197a$. The amount of change is $2.197a - a$ or $1.197a$, and the original amount is a. Dividing the two and multiplying the result by 100, we obtain the answer, 119.7%.

8 Gold.

- Let us call the scarf's original price x. After using one 20% off coupon, the scarf's price is $0.8(x)$.

- Adding another 20% off on top of that, the scarf's price is $0.8(0.8)(x)$. Finally, the last coupon causes the scarf's price to be $0.8(0.8)(0.8)(x)$.

- This expression can be simplified to $0.512x$, or 51.2% of x. 51.2% represents a discount of $100 - 51.2 = 48.8\%$.

9 Silver.

- $32\% = \dfrac{32}{100}$ or $\dfrac{8}{25}$ as a fraction. $\dfrac{8}{25}$ of $4xy$ means $\dfrac{8}{25} \times 4xy$, which can be simplified to $\dfrac{32}{25}xy$.

10 Silver.

- $25 \times \dfrac{7}{5} = 35$. The amount of increase from 25 to 35 is 10, and the original amount is 25. Therefore, the percent increase is $\dfrac{10}{25} \times 100$ or 40%.

11 Silver.

- 10% of 4000 is $\dfrac{1}{10} \times 4000$ or 400. Therefore, this account yields a $400 gain each year for the company, since simple interest always acts on the original amount of money.

- After four years, this will become 400×4 or $1600.

12 Gold.

- After one month, the interest will be 5% of $40 or $2.

- The interest at two months will be 5% of $42 or $2.10.

- The interest at three months will be 5% of $44.10 or $2.21 rounded to the nearest cent.

- When dealing with money, always round to the nearest cent. The total amount of money that the family will have to pay back is $40 + $2 + $2.10 + $2.21 or $46.31.

Part 14: Mixtures

1 Silver.

- We have 5 mL of a solution that is 50% or $\frac{1}{2}$ acid. We want this solution to be 25% or $\frac{1}{4}$ acid.

- Already in the solution are $\frac{1}{2}$ of 5 or 2.5 mL of acid. No matter how much pure water we add to the solution, this amount of acid won't change.

- Therefore, we can set up the equation $2.5/(5 + x) = \frac{1}{4}$ where x is the amount of pure water the scientist is going to add. Solving this equation through cross multiplication, we find that $x = 5$ mL.

2 Gold.

- The scientist's solution so far contains $\frac{20}{100} \times 30$ or 6 mL of acid. Let us call the amount of the second solution that the scientist is going to add x.

- 10% of the second solution is water, so 90% of x is going to be the amount of acid in these x mL. Therefore, $6 + \frac{9}{10}x$ is going to be the total acid content of the final solution.

- The total amount of the final solution will be $30 + x$. We want the ratio of acid content to the total solution to be 30% or $\frac{3}{10}$, so we can set up the equation $\left(6 + \frac{9}{10}x\right)/(30 + x) = \frac{3}{10}$.

- Solving this equation, we find that $x = 5$ mL.

3 Silver.

- If the 20 mL solution is 50% acid, it contains 10 mL of acid.

- After adding x mL of pure water, the total amount of solution will be $20 + x$, but the amount of acid will stay at 10 mL.

- We want the solution to be 10% or $\frac{1}{10}$ acid, so we set up the equation $\frac{10}{20 + x} = \frac{1}{10}$. Solving, we find that $x = 80$ mL.

4 Gold.

- Let us call the number of cookies in the first jar x. We are given that the number of cookies in the second jar is $\frac{1}{2}x$.
- After the second jar's contents are dumped into the first jar, the total amount of cookies is $\frac{3}{2}x$.
- In the first jar, 30% of the x cookies, or $\frac{3}{10}x$ cookies, were yellow. In the second jar, 70% of the $\frac{1}{2}x$ cookies, or $\frac{7}{10} \times \frac{1}{2}x = \frac{7}{20}x$ cookies, were yellow.
- After the dumping, $\frac{3}{10}x + \frac{7}{20}x$ or $\frac{13}{20}x$ cookies are yellow.
- $\frac{13}{20}x / \frac{3}{2}x$ is $\frac{26}{60}$ or 43.3% to the nearest tenth.

5 Silver.

- The liquid in the pot contains 25% of 40 or 10 gallon of mud. After adding x gallon of pure water, the total amount of liquid will be $40 + x$, but the amount of mud will remain at 10 gallon.
- We want the final liquid to be 20% or $\frac{1}{5}$ mud, so we can set up the equation $\frac{\cdot 10}{40 + x} = \frac{1}{5}$.
- Solving this equation, we find that $x = 10$ gallon.

Part 15: Polynomial Expansions

1 Gold.

- To obtain a^2 and $\frac{1}{a^2}$ in the same expression out of $a + \frac{1}{a}$, we square it. We raise both sides of the equation we are given to the second power, obtaining $a^2 + 2(a)\left(\frac{1}{a}\right) + \frac{1}{a^2} = 144$.
- $2(a)\left(\frac{1}{a}\right)$ can be simplified to just 2, so our new equation is $a^2 + 2 + \frac{1}{a^2} = 144$.

- Subtracting 2 from both sides, we find that $a^2 + \dfrac{1}{a^2} = 142$.

2 Bronze.

- In this problem, we multiply the a by both the b and the $+3$, and then add the terms.

- Doing so, we obtain our answer $ab + 3a$.

3 Silver.

- Our strategy in this problem will be to expand $(2n + 3)(n + 1)$ first, and then take that trinomial and multiply it by $(3n + 2)$.

- To expand $(2n + 3)(n + 1)$, we first have to multiply the $2n$ and the n, which gives us $2n^2$.

- Then we multiply the $2n$ and the 1, leaving us with $2n$.

- Moving on to the three, we multiply the 3 and the n, obtaining $3n$, and then the 3 and the 1, obtaining 3.

- Adding all of these yields $2n^2 + 5n + 3$.

- Now, we have to multiply this by $(3n + 2)$. Multiplying each term in the first trinomial by each term in the second and then adding the results, we obtain the long sum $6n^3 + 4n^2 + 15n^2 + 10n + 9n + 6$.

- Combining like terms, we receive our answer: $6n^3 + 19n^2 + 19n + 6$.

4 Silver.

- The sixth row of Pascal's triangle is 1 6 15 20 15 6 1. The highest number in this row is 20, so the highest coefficient in $(m + n)^6$ is 20.

5 Bronze.

- Adding the two a's to make $2a$, we obtain the expression $a^2 + 2a$.

- This is a binomial, because it is the sum of two terms and it cannot be simplified any further.

6 Bronze.

- $(a + b)^2$ can also be written as $(a + b)(a + b)$.

- To expand this, we must multiply the a's by each other, then the a by the b, then the b by the a, and lastly the two b's. Then, we add the results.

- Doing so, we obtain $a^2 + ab + ba + b^2$. ba is the same thing as ab, so we can add this and the first ab to make $2ab$. Now we have our answer, $a^2 + 2ab + b^2$.

7 Silver.

- The fourth row of Pascal's triangle goes as follows: 1 4 6 4 1. This sequence of numbers represents the coefficients of $(m + n)^4$ in standard order.

- Every term in this expansion must have its powers add up to 4, so its terms, without coefficients, are m^4, m^3n, m^2n^2, mn^3, and n^4 in standard order.

- Now we add the necessary coefficients. The term m^4 has coefficient 1, m^3n has coefficient 4, m^2n^2 has coefficient 6, mn^3 has coefficient 4, and n^4 has coefficient 1. Adding these terms yields the answer, $m^4 + 4m^3n + 6m^2n^2 + 4mn^3 + n^4$.

8 Silver.

- Notice that this expression is in the form $a^2 + 2ab + b^2$, where a equals 29 and b equals 11.

- $a^2 + 2ab + b^2$ is the expansion of $(a + b)^2$.

- Hence the expression in the problem equals $(29 + 11)^2$. $29 + 11 = 40$, and 40^2 is easily recognizable as 1600.

9 Bronze.

- $(a + b)^2$ can be expanded into $a^2 + 2ab + b^2$.

- $(a - b)^2$, or $(a - b)(a - b)$, can be expanded into $a^2 - ab - ab + b^2$, or $a^2 - 2ab + b^2$.

- We now have the expression $(a^2 + 2ab + b^2) - (a^2 - 2ab + b^2)$. Distributing the minus sign across the parentheses yields $a^2 + 2ab + b^2 - a^2 + 2ab - b^2$.

- The a^2's and the b^2's cancel, leaving us with $4ab$.

10 Silver.

- Do not get thrown off by the large number of variables. Each pair of variables within the binomials is still just one term.
- First, we multiply the *ab* by the *ef*, obtaining *abef*.
- Next, we multiply the *ab* by the *gh*, obtaining *abgh*.
- After this, we multiply the *cd* by the *ef*, obtaining *cdef*.
- Lastly, we multiply the *cd* by the *gh*, obtaining *cdgh*.
- The final answer is the sum of these products, $abef + abgh + cdef + cdgh$.

11 Bronze.

- To make this problem easier to solve, we will distribute the (13×14) across the parentheses before we do anything else.
- This leaves us with $\frac{2}{13}(13 \times 14) + \frac{3}{14}(13 \times 14)$. In the first part of the sum, the 13s cancel, leaving just 2×14 or 28. In the second part, the 14s cancel, leaving 3×13 or 39.
- $28 + 39 = 67$.

12 Silver.

- In this problem, we substitute y with $2y$ in the general binomial expansion.
- The third row of Pascal's Triangle is 1 3 3 1. Therefore, this expansion equals $x^3 + 3x^2(2y) + 3x(2y)^2 + (2y)^3$, if we go in standard order, each time decreasing the power of x by 1 and increasing the power of $2y$ by 1.
- If we simplify this expression, we obtain $x^3 + 6x^2y + 12xy^2 + 8y^3$.

13 Gold.

- After the first year, the savings account will gain $p\%$ of d dollars, or $\frac{p}{100} \times d = \frac{pd}{100}$ dollars. This account will now be worth $d + \frac{pd}{100}$ dollars.

- After the second year, $\dfrac{p}{100} \times \left(d + \dfrac{pd}{100} \right)$ dollars will be added.
  This is $\dfrac{pd}{100} + \dfrac{p^2 d}{10000}$ when simplified. The account is now
  worth $d + \dfrac{pd}{100} + \dfrac{pd}{100} + \dfrac{p^2 d}{10000}$ or $d + \dfrac{pd}{50} + \dfrac{p^2 d}{10000}$ dollars.

- After the third year, $\dfrac{p}{100} \times \left(d + \dfrac{pd}{50} + \dfrac{p^2 d}{10000} \right)$ dollars will
  be added. This is $\dfrac{pd}{100} + \dfrac{p^2 d}{5000} + \dfrac{p^3 d}{1000000}$ when simplified.

- The account is now worth $d + \dfrac{pd}{50} + \dfrac{p^2 d}{10000} + \dfrac{pd}{100} + \dfrac{p^2 d}{5000} +$
  $\dfrac{p^3 d}{1000000}$ or $d + \dfrac{3pd}{100} + \dfrac{3p^2 d}{10000} + \dfrac{p^3 d}{1000000}$ dollars when
  simplified.

14 Platinum.

- The fifth row of Pascal's Triangle is 1 5 10 10 5 1. Therefore,
  $(x + y)^5 = x^5 + 5x^4 y + 10x^3 y^2 + 10x^2 y^3 + 5xy^4 + y^5$.

- If we substitute $3xy$ and 4 in for x and y respectively, we can
  expand $(3xy + 4)^5$ into $(3xy)^5 + 5(3xy)^4(4) + 10(3xy)^3(4)^2 +$
  $10(3xy)^2(4)^3 + 5(3xy)(4)^4 + (4)^5$

- $(3xy)^5 = 243x^5 y^5$

- $5(3xy)^4(4) = 5(81x^4 y^4)(4) = 1620x^4 y^4$

- $10(3xy)^3(4)^2 = 10(27x^3 y^3)(16) = 4320x^3 y^3 10$

- $10(3xy)^2(4)^3 = 10(3xy)^2(4)^3 = 10(9x^2 y^2)(64) = 5760x^2 y^2$.

- $5(3xy)(4)^4 = 15xy(256) = 3840xy$

- $4^5 = 1024$.

- The sum of these is: $243x^5 y^5 + 1620x^4 y^4 + 4320x^3 y^3 +$
  $5760x^2 y^2 + 3840xy + 1024$.

15 Bronze.

- $3x \times x = 3x^2$, $3x \times z = 3xz$, $2y \times x = 2xy$, $2y \times z = 2yz$,
  $4z^2 \times x = 4xz^2$, $4z^2 \times z = 4z^3$.

- Therefore, our answer is $3x^2 + 3xz + 2xy + 2yz + 4xz^2 + 4z^3$.

Part 16: Equivalent Expressions

1 Silver.

- Remember, terms with x^2 cannot be combined through addition or subtraction with terms with any other power of x. We are solving for an expression in terms of x and x^2, so we must leave them alone.

- $ax^2 + bx^2$ must correspond to $7x^2$, $ax - bx$ must correspond to $4x$, and c must correspond to 8. We instantly know that $c = 8$. Finding a and b will take more work.

- If $ax^2 + bx^2 = 7x^2$, $a + b = 7$. Similarly, we find that $a - b = 4$.

- We now have a solvable system of equations: $a + b = 7$ and $a - b = 4$. Using elimination, $(a+b)+(a-b) = 7+4$. Simplifying yields $2a = 11$, and therefore $a = 11/2$. Substituting the value of a back into $a + b = 7$ or $a - b = 4$, we find that $b = 3/2$.

2 Silver.

- Since we are solving for an expression that is in terms of x, x cannot take on any specific value. Therefore, $2ax + 3bx$ must correspond to $22x$ and $3a + 2b$ must correspond to 34.

- Since $2ax+3bx = 22x$, $2a+3b = 24$. This along with $3a+2b = 34$ forms a solvable system of equations.

- One way to find $a+b$ would be to solve for both a and b and then find their sum. However, adding the two equations yields $5a + 5b = 58$. We can rewrite this as $5(a + b) = 58$, and therefore $a + b = 58/5$.

- This trick is a bit difficult to come up with on the spot. However, solving for a and b individually works just as well, only it takes more time.

Part 17: Factoring

1 Bronze.

- The first thing we do is factor out $2x$, since every term is divisible by it. We factor the expression into $2x(x^3 + 2x + 6)$

- There is no way to further factor this, so we are finished.

2 Bronze.

- The numerator of this fraction, $3x + 9$, can be factored into $3(x + 3)$, since both $3x$ and 9 can be divided evenly by 3.

- Since $x + 3$ is in both the numerator and the denominator, both instances can be canceled. This leaves us with just 3, except when $x = -3$.

3 Silver.

- $2x^2 + 6x$ can be factored into $2x(x + 3)$, since $2x$ divides both terms evenly. If either $2x$ or $(x + 3)$ is 0, the whole term will equal 0 regardless of what the other's value is.

- Therefore, the two equations $2x = 0$ and $x + 3 = 0$ determine our answers. Solving the equations, we obtain $x = 0$ and $x = -3$.

4 Silver.

- This trinomial will be factored into the product of two binomials: $(ex + f)(gx + h)$. According to our factoring rules, $e \times g$ has to equal 1, $f \times h$ has to equal 9, and $eh + fg$ has to equal 6.

- Since the coefficient of x^2 in the trinomial is 1, both e and g have to be 1 as well, since no other whole numbers multiply to 1. We can make f and h 1 and 9 in either order or 3 and 3.

- If f and h are 9 and 1 in either order, $eh + fg$ will turn out to be 10, but if both e and f are 3, $eh + fg$ does in fact equal 6. Therefore, our answer is $(x + 3)(x + 3)$ or $(x + 3)^2$.

5 Silver.

- We already know that this trinomial can be factored into $(x + 3)(x + 3)$ from the previous problem. We also know that if either of these binomials is equivalent to 0, the whole expression will equal 0 and the equation will be solved.

- In both of these binomials, $x = -3$ is the solution. Therefore, $x = -3$ is the only solution. Not all quadratics have two solutions.

6 Silver.

- The first thing we can do is factor the trinomial into $2(3x^2 + 10x + 8)$. Now we move on to factoring the simpler trinomial into two binomials.

- In the factorization $(ex + f)(gx + h)$, $e \times g$ must equal 3, $f \times h$ must equal 8, and $eh + fg$ must equal 10.

- e and g can either be 3 and 1 or 1 and 3 respectively. f and h can either be 8 and 1, 4 and 2, 2 and 4, or 1 and 8 respectively.

- Testing every possible combination, we find that when e is 3, g is 1, f is 4, and h is 2, $eh + fg = 20$, so the factorization is $(3x + 4)(x + 2)$.

- Our final answer is $2(3x + 4)(x + 2)$.

7 Silver.

- The first step in solving the equation is to factor $x^2 + 5x + 4$.

- If we use the format $(ex + f)(gx + h)$, both e and g have to equal 1, since no other whole numbers multiply to 1, and f and h have to be 1 and 4, 2 and 2, or 4 and 1 respectively.

- When f and h are 4 and 1 in either order, $eh + fg$ equals 5, so our factorization is $(x + 1)(x + 4)$. Now, our equation is $(x + 1)(x + 4) = 0$.

- If either $(x + 1)$ or $(x + 4)$ is 0, no matter what the other binomial comes out to be, the product will equal 0. Therefore, our two solutions are the solution of $x + 1 = 0$ and the solution of $x + 4 = 0$, which are $x = -1$ and -4.

8 Silver.

- Let us call the mystery number x. The expression denoting the changes made to the variable is $6x + 9 + x^2$, or $x^2 + 6x + 9$. We have that this equals 4, so our equation is $x^2 + 6x + 9 = 4$.

- We look to receive an expression equivalent to 0, so we subtract 4 from both sides to obtain $x^2 + 6x + 5 = 0$. Now, we must factor the trinomial.

- In the factorization $(ex + f)(gx + h)$, both e and g must equal 1, and f and h have to multiply to 5, so they are either 1 and 5 or 5 and 1 respectively.

- If f and h are 1 and 5, then $eh + fg = 6$, so our equation now is $(x + 1)(x + 5) = 0$.

- The equation will hold true if $x + 1 = 0$ or if $x + 5 = 0$, so $x = -1$ or -5. Since -1 is the only one of the two greater than -3, it is our answer.

9 Silver.

- To find the value of $x - y$, we must obtain an $x - y$ in our equation.

- To do this, we subtract 2y and add 32 to both sides of the equation. This leaves us with $2x - 2y = 32$.

- If we factor out 2 from $2x - 2y$, we obtain the equation $2(x - y) = 32$. Dividing both sides by 2, we find that $x - y = 16$.

10 Silver.

- Let us try to factor the numerator, $3x^2 + 17x + 10$. The product of the two resulting binomials will be in the format $(ex + f)(gx + h)$.

- The only two whole numbers that multiply to 3 are 1 and 3, so we will set e equal to 1 and g equal to 3. (If we set e as 3 and g as 1, we should get the same pairs of binomials, just in reverse order.)

- f and h must multiply to 10, so they are either 1 and 10, 2 and 5, 5 and 2, or 10 and 1 respectively. After testing the value of $eh + fg$ for each one, we find that when f is 5 and h is 2, $eh + fg$ does equal 17. Therefore, $3x^2 + 17x + 10$ factored is $(x + 5)(3x + 2)$.

- Our expression now is $(x + 5)(3x + 2)/(x + 5)$. Since any real number divided by itself is 1, the $(x + 5)$s on the top and bottom cancel. Therefore, our answer is $1(3x + 2)$ or $(3x + 2)$, except when $x = -5$.

11 Silver.

- Let us call the lengths of the four bars a, b, c, and d, respectively. After the increase of 24%, these four bars will have lengths of $1.24a$, $1.24b$, $1.24c$, and $1.24d$, respectively.

- The original sum of the lengths of the four bars was $a+b+c+d$, and now the sum is $1.24a + 1.24b + 1.24c + 1.24d$. This new sum can be factored into $1.24(a + b + c + d)$. It is now easy to see that the sum also increased by 24%.

- This problem serves as an example to the fact that if you increase or decrease each thing in a sum by the same percent, the sum increases or decreases by that percent as well.

12 Gold.

- This expression is in the form $ax^2 + bx + c$, except instead of x, the variable is x^2. x^2 can be treated as a single variable.

- The expression $x^4 + 4x^2 + 4$ can be written as $(x^2)^2 + 4(x^2) + 4$. This can be factored into $(e(x^2) + f)(g(x^2) + h)$ for some values of e, f, g, and h.

- Since e and g multiply to 1, e and g both have to equal 1, because no other pair of whole numbers multiplies to 1. Since f and h multiply to 4, they are either 1 and 4, 2 and 2, or 4 and 1 respectively.

- When f and h both equal 2, $eh + fg = 4$, so our factorization is $(x^2 + 2)(x^2 + 2)$, or $(x^2 + 2)^2$.

13 Gold.

- The square of the sum of x, y, and z can be written as $(x + y + z)^2$.

- What we do now is expand this expression. $(x + y + z)^2 = (x + y + z)(x + y + z) = x^2 + xy + xz + xy + y^2 + yz + xz + yz + z^2 = x^2 + y^2 + z^2 + 2xy + 2xz + 2yz$.

- We have that the sum of the squares of x, y, and z, or $x^2 + y^2 + z^2$, is 15, and $xy + yz + xz = 10$. $2xy + 2xz + 2yz = 2(xy + yz + xz)$, so $2xy + 2xz + 2yz = 2(10)$ or 20.

- Substituting the values of these two expressions into the previous expression, we find that $(x + y + z)^2 = 15 + 20 = 35$.

14 Platinum.

- Let us try analyzing what we are being asked to find before we factor this complex trinomial. Even if it cannot be factored into nice numbers, any trinomial can be factored into $(ex + f)(gx + h)$. Since the coefficient of x^2 is 1, we can assume that both e and g equal 1. Our factorization is now $(x + f)(x + h)$.

- The roots to this equation are determined by $x + f = 0$ and $x + h = 0$. Therefore, the roots are $-f$ and $-h$. The sum of these roots is $-f + (-h)$, which can be written as $-1(f + h)$.

- Recall that $eh + fg$ must equal the middle term of the trinomial, or 987. In this case, both e and g are 1, so $eh + fg$ is just $h + f$.

- $h + f = 987$, so $-1(f + h) = -987$.

Part 18: Special Factorizations

1 Silver.

- The way to remove $\sqrt{10}$ from the denominator is to multiply both the numerator and denominator of the fraction by $\sqrt{10}$. This will not change its value, because we are essentially multiplying by 1.

- $\sqrt{10} \times \sqrt{10} = 10$, and $10 \times \sqrt{10} = 10\sqrt{10}$. Therefore, this fraction equals $\dfrac{10\sqrt{10}}{10}$. 10 cancels on both top and bottom, leaving just $\dfrac{\sqrt{10}}{1}$ or $\sqrt{10}$.

2 Silver.

- To rationalize the denominator, we multiply by $\dfrac{\sqrt{12}}{\sqrt{12}}$. This does not change the term's value, as we are essentially multiplying by 1.

- $\dfrac{\sqrt{6}}{\sqrt{12}} \times \dfrac{\sqrt{12}}{\sqrt{12}} = \dfrac{\sqrt{72}}{12} = \dfrac{6\sqrt{2}}{12} = \dfrac{\sqrt{2}}{2}$.

3 Silver.

- This expression is in the form $a^2 - b^2$. Since $a^2 - b^2 = (a+b)(a-b)$, $51^2 - 49^2 = (51+49)(51-49)$.

- $51 + 49 = 100$, and $51 - 49 = 2$. $100 \times 2 = 200$.

4 Gold.

- To remove both $\sqrt{6}$ and $\sqrt{10}$ from the denominator in a single step, we use the $a^2 - b^2$ factorization. The denominator is currently in the form $a + b$, so we have to multiply by $a - b$ to obtain $a^2 - b^2$.

- Multiplying both the numerator and denominator of the fraction by $\sqrt{6} - \sqrt{10}$ yields $\dfrac{\sqrt{6} - \sqrt{10}}{6 - 10}$, or $\dfrac{\sqrt{6} - \sqrt{10}}{-4}$. This can be simplified to $\dfrac{\sqrt{10} - \sqrt{6}}{4}$.

5 Gold.

- The denominator is in the form $a - b$. We must multiply by $a + b$ to square both a and b.

- Multiplying both the numerator and denominator by $\sqrt{10} + \sqrt{6}$ yields $\dfrac{(\sqrt{10} + \sqrt{6})(\sqrt{10} + \sqrt{6})}{(\sqrt{10} - \sqrt{6})(\sqrt{10} + \sqrt{6})}$.

- Expanding the numerator yields $10 + \sqrt{60} + \sqrt{60} + 6$, which can be simplified to $16 + 2\sqrt{60}$. Expanding the denominator yields $10 - 6$ or 4. $2\sqrt{60}$ can be simplified to $4\sqrt{15}$, so the original fraction is equivalent to $\dfrac{16 + 4\sqrt{15}}{4}$.

- We split this up into $\dfrac{16}{4} + \dfrac{4\sqrt{15}}{4}$ to obtain $4 + \sqrt{15}$ as the final answer.

6 Silver.

- $a^3 + b^3$ can be expanded into $(a + b)(a^2 - ab + b^2)$. We already know the values of $(a + b)$ and ab, so all we need to find is the value of $a^2 + b^2$. (Remember, $a^2 - ab + b^2 = a^2 + b^2 - ab$).

- Since $(a + b) = 3$, $(a + b)^2$ or $a^2 + 2ab + b^2 = 9$. Replacing ab with 4 yields $a^2 + 8 + b^2 = 9$, and subtracting 8 from both sides yields $a^2 + b^2 = 1$.

- We now have all the information we need to find $(a + b)(a^2 - ab + b^2)$. We replace $(a + b)$ with 3, $-ab$ with -4, and $a^2 + b^2$ with 1. Doing so leaves us with $3(1 - 4)$, which equals -9.

7 Silver.

- $a^3 - b^3$ can be expanded into $(a - b)(a^2 + ab + b^2)$. We already know the values of $(a - b)$ and ab, so all we need to find is the value of $a^2 + b^2$.

- Since $(a - b) = 10$, $(a - b)^2$ or $a^2 - 2ab + b^2 = 10^2$ or 100. Replacing ab with 6 yields $a^2 - 12 + b^2 = 100$, and adding 12 to both sides yields $a^2 + b^2 = 112$.

- We now have all the information we need to find $(a - b)(a^2 + ab + b^2)$. We replace $(a - b)$ with 10, ab with 6, and $a^2 + b^2$

with 112. Doing so leaves us with $10(112 + 6)$, which equals 1180.

8 Gold.

- $a^2 + 2ab + b^2$ can be factored into $(a + b)^2$. We now have $(a + b)^2 - c^2$.
- This is an expression in the form $x^2 - y^2$, except x in this case is $(a + b)$.
- $x^2 - y^2$ can be factored into $(x + y)(x - y)$, so $(a + b)^2 - c^2$ can be factored into $((a+b)+c)((a+b)-c)$, which can be rewritten as $(a + b + c)(a + b - c)$.

Part 19: The Quadratic Formula

1 Silver.

- Using the quadratic formula, we obtain the two solutions $\dfrac{-9 + \sqrt{9^2 - 4(2)(7)}}{2(2)}$ and $\dfrac{-9 - \sqrt{9^2 - 4(2)(7)}}{2(2)}$.
- The first solution simplifies to $\dfrac{-9 + \sqrt{25}}{4}$ and the second simplifies to $\dfrac{-9 - \sqrt{25}}{4}$. Therefore, the final answer is $a = -1$ and $a = -7/2$.

2 Silver.

- Using the quadratic formula, we obtain the two solutions

$$\frac{-(-1) + \sqrt{(-1)^2 - 4(3)(-12)}}{2(3)}$$

and

$$\frac{-(-1) - \sqrt{(-1)^2 - 4(3)(-12)}}{2(3)}.$$

- $(-1)^2 - 4(3)(-12) = 1 - (-144) = 145$. Therefore, the final answer is $a = \dfrac{1 + \sqrt{145}}{6}$ and $\dfrac{1 - \sqrt{145}}{6}$.

3 Silver.

- Using the quadratic formula, we obtain the two solutions

$$\frac{-(-2) + \sqrt{(-2)^2 - 4(1)(31)}}{2(1)}$$

and

$$\frac{-(-2) + \sqrt{(-2)^2 - 4(1)(31)}}{2(1)}.$$

- $(-2)^2 - 4(1)(31)$ is negative, so taking its square root cannot yield a real result. Therefore, this equation has no solution.

4 Silver.

- We need to have a quadratic expression equivalent to 0 in order to use the quadratic formula.

- To get there, we add 8 to both sides. Doing so yields the new equation $5x^2 + 10x + 5 = 0$. The left side of the equation can be factored into $5(x^2 + 2x + 1)$, and we can simplify by dividing both sides by 5.

- Doing so leaves us with $x^2 + 2x + 1 = 0$. Using the quadratic formula, the two solutions to this equation are $\frac{-2 + \sqrt{2^2 - 4(1)(1)}}{2(1)}$ and $\frac{-2 - \sqrt{2^2 - 4(1)(1)}}{2(1)}$. Both turn out to be $x = -1$.

5 Silver.

- First, we subtract 5 from both sides to make the right side of the equation 0. This leaves us with $3m^2 - m\sqrt{7} - 27 = 0$.

- Using the quadratic formula, we find that the two solutions are

$$\frac{-(-\sqrt{7}) + \sqrt{(-\sqrt{7})^2 - 4(3)(-27)}}{2(3)}$$

and

$$\frac{-(-\sqrt{7}) - \sqrt{(-\sqrt{7})^2 - 4(3)(-27)}}{2(3)}.$$

These simplify to $m = \frac{\sqrt{7} + \sqrt{331}}{6}$ and $m = \frac{\sqrt{7} - \sqrt{331}}{6}$, respectively.

6 Silver.

- Simplifying the right side of the equation through addition and subtraction yields $4a^3 - 15a^2 + 21a + 24$.

- There is one $4a^3$ on each side of the equation, so we can subtract $4a^3$ from both sides to obtain $-21a^2 - 10a + 23 = -15a^2 + 21a + 24$. As of right now, this equation cannot be solved with the quadratic formula.

- Let us add $15a^2$ and subtract $21a$ and 24 from both sides. Doing so yields $-6a^2 - 31a - 1 = 0$. Using the quadratic formula, the two solutions are $a = \dfrac{-(-31) + \sqrt{(-31)^2 - 4(-6)(-1)}}{2(-6)}$ and $a = \dfrac{-(-31) - \sqrt{(-31)^2 - 4(-6)(-1)}}{2(-6)}$.

- These simplify to $a = \dfrac{-31 + \sqrt{937}}{12}$ and $\dfrac{-31 - \sqrt{937}}{12}$, respectively.

7 Gold.

- To solve this equation, we must remove c from the denominator of the first fraction on the left side. To do this, we multiply both sides by c. This will cancel the fraction's denominator.

- $c\left(\dfrac{1}{c} + c\right) = (4)c$ simplifies to $1 + c^2 = 4c$. Subtracting $4c$ from both sides yields the solvable quadratic equation $c^2 - 4c + 1 = 0$. Using the quadratic formula, we obtain the two solutions $c = \dfrac{-(-4) + \sqrt{(-4)^2 - 4(1)(1)}}{2(1)}$ and $c = \dfrac{-(-4) - \sqrt{(-4)^2 - 4(1)(1)}}{2(1)}$.

- These simplify to $c = \dfrac{4 + \sqrt{12}}{2}$ and $c = \dfrac{4 - \sqrt{12}}{2}$. They further simplify to $c = \dfrac{4 + 2\sqrt{3}}{2}$ and $\dfrac{4 - 2\sqrt{3}}{2}$, which simplify even further to $2 + \sqrt{3}$ and $2 - \sqrt{3}$.

8 Gold.

- First and foremost, we must take the denominators out of the picture.

- We multiply both sides of the equation by $(x + 5)$ and then by $(2x + 1)$. Doing so yields $(2x + 3)(2x + 1) = (3x + 2)(x + 5)$. Expanding yields $4x^2 + 8x + 3 = 3x^2 + 17x + 10$.

- Subtracting $3x^2$, $17x$, and 10 from both sides leaves us with $x^2 - 9x - 7 = 0$. Using the quadratic formula to solve for x, we find that the two solutions are

$$x = \frac{-(-9) + \sqrt{(-9)^2 - 4(1)(-7)}}{2(1)}$$

and

$$x = \frac{-(-9) - \sqrt{(-9)^2 - 4(1)(-7)}}{2(1)}.$$

- These simplify to $x = \dfrac{9 + \sqrt{109}}{2}$ and $x = \dfrac{9 - \sqrt{109}}{2}$, respectively.

9 Gold.

- Let us call the two numbers x and y. We have that $x + y = 17$ and $xy = 65$.

- Let us use substitution to solve this system of equations. Since $x + y = 17$, $y = 17 - x$. Replacing y with this expression in the second equation yields $x(17 - x) = 65$, which can be expanded into $17x - x^2 = 65$.

- This is a quadratic equation. If we rearrange it into $x^2 - 17x + 65 = 0$, we can use the quadratic formula to solve for x. Doing so, we find that the two solutions for x are $\dfrac{17 + \sqrt{29}}{2}$ and $\dfrac{17 - \sqrt{29}}{2}$.

- Now, we substitute these two values back into $x + y = 17$. Subtracting $\dfrac{17 + \sqrt{29}}{2}$ from both sides in the equation $\dfrac{17 + \sqrt{29}}{2} + y = 17$ yields $y = 17 - \dfrac{17 + \sqrt{29}}{2}$.

- Using a common denominator, we can rewrite this as $y = \dfrac{34 - (17 + \sqrt{29})}{2}$. Distributing the negative across the parentheses, we find that $y = \dfrac{17 - \sqrt{29}}{2}$.

- However, we still have one more value of x to account for. Subtracting $\dfrac{17 - \sqrt{29}}{2}$ from both sides in the equation $\dfrac{17 - \sqrt{29}}{2} + y = 17$ yields $y = 17 - \dfrac{17 - \sqrt{29}}{2}$. Using a common denominator, we can rewrite this as $y = \dfrac{34 - (17 - \sqrt{29})}{2}$. Distributing the negative across the parentheses, we find that $y = \dfrac{17 + \sqrt{29}}{2}$.

- Our two solutions are made up of the same numbers. Therefore, our answer is just $\dfrac{17 + \sqrt{29}}{2}$ and $\dfrac{17 - \sqrt{29}}{2}$.

10 Silver.

- Since the quadratic equation has one real solution, $\sqrt{b^2 - 4ac} = 0$. If this were to be false, the equation would either have no real solutions or two real solutions.

- We are given that $a = 2$ and $c = 10$. The equation $\sqrt{b^2 - 4(2)(10)} = 0$ can be used to find b.

- The only square root of 0 is 0, so $b^2 - 4(2)(10)$ or $b^2 - 80$ is equivalent to 0.

- It follows that there are two possible values of b: $b = \sqrt{80}$ and $b = -\sqrt{80}$. Simplifying $\sqrt{80}$ to $4\sqrt{5}$ yields $b = 4\sqrt{5}$ and $b = -4\sqrt{5}$.

11 Silver.

- Subtracting x and then 4 from both sides yields $11x^2 + 21x + 11 = 0$. To find the number of real solutions to this equation, we examine the value of $\sqrt{b^2 - 4ac}$.

- In this case, $\sqrt{b^2 - 4ac} = \sqrt{21^2 - 4(11)(11)} = \sqrt{-43}$. Square roots of negatives are not real numbers, so the quadratic equation has no solutions.

Part 20: Solving for Whole Expressions

1 Silver.

- Substituting y for $\frac{12}{13}z$ in the equation $x = \frac{1}{2}y$, we obtain $x = \frac{6}{13}z$.

- The problem asks for the fraction of z that is x. This is a little bit cryptic, but we know that the desired solution written as an equation is $x =$ fraction of z. Therefore, the answer is $\frac{6}{13}$.

2 Silver.

- Adding the two equations yields $7x + 7y = 28$. Factoring the left side yields $7(x + y) = 28$, and therefore $x + y = 4$.

- The problem asks us to find $3x + 3y$. This is equivalent to $3(x + y)$. Replacing $x + y$ with 4, we find that $3x + 3y = 12$.

3 Gold.

- Another way to write $-1\frac{1}{2}y + \frac{1}{2}x$ is $\frac{1}{2}x - \frac{3}{2}y$. If we factor out $\frac{1}{2}$ from this expression, we obtain $\frac{1}{2}(x - 3y)$.

- If we can find the value of $x - 3y$ from the equation we are given, all we have to do to is multiply by $\frac{1}{2}$ and we receive our answer.

- $5x - 15y$ can be factored into $5(x - 3y)$, leaving us with $5(x - 3y) = 42$. Dividing both sides of the equation by 5, we find that $x - 3y = \frac{42}{5}$.

- $\frac{42}{5} \times \frac{1}{2} = \frac{21}{5}$, our answer.

4 Gold.

- Adding equations does not seem to get us anywhere. But what about subtracting them (the same thing as multiplying both sides of the second equation by -1 and then adding the equations).

- Subtracting these two equations yields $(3x + 2y + 5z) - (x + 4y + 7z) = 21 - 30$. Simplifying, we obtain $2x - 2y - 2z = -9$.
- $2x - 2y - 2z$ can be factored into $2(x - y - z)$. What we are left with now is $2(x - y - z) = -9$, and therefore $x - y - z = -\dfrac{9}{2}$.

5 Platinum.

- Recognize that $x^3 + 3x^2 + 3x + 1$ can be factored into $(x + 1)^3$ through Pascal's Triangle.
- In this equation, we have an $x^3 + 3x^2 + 3x + 5$, but if this was $x^3 + 3x^2 + 3x + 1$, the problem would be much easier. Therefore, we subtract 4 from both sides. This leaves us with $x^3 + 3x^2 + 3x + 1 = 15$, which can be rewritten as $(x + 1)^3 = 15$.
- Taking the cube root of both sides, we find that $x + 1 = \sqrt[3]{15}$, and therefore $x = \sqrt[3]{15} - 1$.

6 Silver.

- Subtracting $5x$ and then adding y to both sides, we obtain $-2x = 25y$. To receive an $\dfrac{x}{y}$ in our equation, we can divide both sides by y.
- This yields $\dfrac{-2x}{y} = 25$, which can be rewritten as $-2\left(\dfrac{x}{y}\right) = 25$. Dividing both sides by -2, we find that $\dfrac{x}{y} = -\dfrac{25}{2}$.

7 Gold.

- Note that the value of xy can be found from $y = \dfrac{13}{x}$ by multiplying both sides by x. Doing this yields $xy = 13$.
- We did this because there is an xy in $xy + yz = 35$. Substituting the value we found into this equation, we obtain $13 + yz = 35$, and therefore $yz = 22$.
- We now have that $yz = 22$ and $y + z = 21$. From $yz = 22$, we can derive that $z = \dfrac{22}{y}$. Plugging this value into the second equation yields $y + \dfrac{22}{y} = 21$.

- If we multiply both sides by y, we will be left with a quadratic equation: $y\left(y + \dfrac{22}{y}\right) = 21y$ can be simplified into $y^2 + 22 = 21y$, which can be rearranged into $y^2 - 21y + 22 = 0$.

- Using the quadratic formula, we find that $y = \dfrac{21 + \sqrt{353}}{2}$ or $\dfrac{21 - \sqrt{353}}{2}$.

8 Gold.

- How do we obtain an $a - c$ with these two equations? One way would be to subtract the two equations (the same thing as multiplying both sides of the second equation by -1 and then adding the equations).

- Doing this yields $(a + b) - (b + c) = 7 - 8$. Distributing the subtraction across the parentheses yields $(a+b)-b-c = 7-8$, which simplifies into $a - c = -1$.

Part 21: Infinite Series

1 Silver.

- The common ratio here is $\dfrac{1}{2}$, and the first term is 8. Therefore, the sum of this infinite geometric series is $8 \,/\, \dfrac{1}{2}$ or 16.

2 Bronze.

- There are two digits being repeated here, so by our "9-trick," this decimal equals $\dfrac{13}{99}$.

3 Silver.

- The common ratio here is $\dfrac{1}{5}$, and the first term is 10. Therefore, the sum of this infinite geometric series is $10/\dfrac{4}{5}$, or $\dfrac{25}{2}$.

4 Platinum.

- Let us set the value of this series to x. The square of this series, or x^2, equals

$$110 - \sqrt{110 - \sqrt{110 - \sqrt{110 - \sqrt{110 - \sqrt{110}}}}} \ldots$$

- After the $110 -$, this series is identical to x. Therefore, $x^2 = 110 - x$.

- This equation can be rewritten into $x^2 + x - 110 = 0$. Solving this equation, we find that $x = 10$.

5 Platinum.

- Let us set the value of this series to x. The reciprocal of this series, or $\frac{1}{x}$, is $5 + \cfrac{1}{5 + \cfrac{1}{5 + \cfrac{1}{5 + 1 \ldots}}}$.

- After the $5 +$, this series is identical to our original series. Therefore, $\frac{1}{x} = 5 + x$. Solving this equation, we find that $x = \frac{\sqrt{29} - 5}{2}$. Note that this series cannot possibly converge to a negative value.

6 Silver.

- This series keeps getting larger and larger forever without approaching any value. Therefore, its sum is infinity.

7 Platinum.

- Let us set the value of this series to x. Our goal is to manipulate this series in a way that yields a repeat of the infinite series and subsequently a way to plug x into itself.

- If we multiply this series by 3, we obtain $\frac{1}{2} + \frac{1}{3} \times \left(\frac{1}{2} + \frac{1}{3} \times \left(\frac{1}{2} + \frac{1}{3} \times \left(\frac{1}{2} \ldots \right) \right) \right)$. If we then subtract $\frac{1}{2}$ from this, what we are left with is simply x. Therefore, $3x - \frac{1}{2} = x$.

- Solving this equation, we find that $x = \frac{1}{4}$.

8 Gold.

- Let us set 0.12555 ... equal to n. If we multiply n by 100, we receive 12.55555 ... .

- It is easy to see that this value equals $12\frac{5}{9}$. We have that $100n = 12\frac{5}{9}$. Converting the left side to an improper fraction and then dividing both sides by 100, we find that $n = \frac{113}{900}$.

- The reciprocal of this is $\frac{900}{113}$.

Part 22: Sets

1 Bronze.

- The set of all the unique elements in either of these sets is $\{a, c, d, f, g\}$.

2 Bronze.

- Both of these sets have an a and a c. Therefore, the intersection of these sets is $\{a, c\}$.

3 Bronze.

- The intersection of $\{a, b, c\}$ and $\{a, d, e\}$ is $\{a\}$, since both of these sets have a in common.

- We are also given that the intersection set equals $\left\{\frac{1}{2}\right\}$, so $\{a\} = \left\{\frac{1}{2}\right\}$ and therefore $a = \frac{1}{2}$.

4 Silver.

- The number of elements in one set plus the number of elements in a second set minus the intersection of the two sets equals the number of sets in their union.

- Our two sets in this case are the set of all the students that own saxophones and the set of all the students that own tubas. We already know that the union of these two sets is $200 - 30 =$

170, since there are 200 total students and 30 of them own neither instrument.

- Let us call the number of students who own both instruments x. The number of students who own only a saxophone is $98 - x$, and the number of students who own only a tuba is $120 - x$.

- $120 - x + 98 - x + x = 170$, since the number of students who play only the tuba plus the number of students who play only the saxophone plus the number of students who play both equals the union of the sets. Solving this equation, we find that $x = 48$ students.

5 Gold.

- Let us call the number of customers who purchased all three items x. $14 - x$ customers purchased burgers and appetizers but not drinks, $21 - x$ customers purchased drinks and appetizers but not burgers, and $20 - x$ customers purchased drinks and burgers but not appetizers.

- The number of customers who purchased only drinks is $60 - (x + (21 - x) + (20 - x)) = 19 + x$, the number of customers who purchased only burgers is $60 - (x + (14 - x) + (20 - x)) = 26 + x$, and the number of customers who purchased only appetizers is $49 - (x + (14 - x) + (21 - x)) = 14 + x$. All of these values should add up to the total number of customers, 120, since we have covered every part of the three-set intersection.

- Therefore, we set up the equation $x + 14 - x + 21 - x + 20 - x + 19 + x + 26 + x + 14 + x = 120$. Simplifying and then solving for x, we find that 6 customers purchased all three items.

Part 23: Mean, Median, Mode and Range

1 Bronze.

- The mean of the set is $\dfrac{a + b}{2}$, since the sum of the elements is $a + b$ and the number of elements is 2.

- This set has an even number of elements and no element exactly in the middle. To find its median, we take the mean of the elements right next to the exact middle. These are a and b. The mean of a and b is $\dfrac{a+b}{2}$. We now see that the set's mean is equivalent to its median.

- Since anything subtracted from itself equals 0, the difference between the median and the mean of the set is 0.

2 Silver.

- The mean of the set is $\dfrac{a+b+c}{3}$.

- The set is ordered; therefore, $a \le b \le c$. Since c has the highest value and a has the lowest value, the range is $c - a$.

- b occupies the middle position, so the median of the set is b.

- We have that $\dfrac{a+b+c}{3} = c - a + b$. $\dfrac{a+b+c}{3}$ can be rewritten as $\dfrac{a}{3} + \dfrac{b}{3} + \dfrac{c}{3}$. Adding and subtracting terms from both sides to get a and its coefficient alone on one side, we obtain $\dfrac{4}{3}a = \dfrac{2}{3}b + \dfrac{2}{3}c$.

- Multiplying both sides by $\dfrac{3}{4}$ to isolate a, we find that $a = \dfrac{1}{2}b + \dfrac{1}{2}c$.

3 Silver.

- If the mean number of marbles among the four jars is 25, the total number of marbles in the four jars is 100. This is because the total number of marbles in the four jars divided by four equals 25 by the definition of mean.

- Another jar is added, so there are five jars in total. The mean number of marbles among the five jars is 30, so the five jars have 30×5 or 150 marbles in total.

- It follows that there are 50 marbles in the fifth jar.

4 Silver.

- The first step is to put the set in order: $\{1, 2, 5, 5, 6, 9, 10, 12, 14\}$.

- To find the mode, we identify the value that occurs most often. This is 5.

- To find the range, we subtract the least value from the greatest. This equals $14 - 1$ or 13.

- The set has nine terms. The median of the (ordered) set is the fifth, which is 6.

5 Gold.

- If the set has a mean of 7, the sum of its 13 elements is 13×7 or 91.

- We are given that 8 is the median, so the seventh element of the set when put in increasing order is 8.

- For one element to be as high as possible, the other elements must be as low as possible, since there is a limit on the sum of the elements.

- 1 is the smallest positive integer, so 1 is the smallest number that can be in the set. We shall therefore set the first 6 elements in this set equal to 1.

- How do we make elements $7 - 12$ as small as possible knowing that the seventh term is 8? None of these elements can be lower than 8, because the median of the set *in increasing order* is 8.

- We set the 7th through 12th terms equivalent to 8, and now the sum of the first 12 elements is $1 + 1 + 1 + 1 + 1 + 1 + 8 + 8 + 8 + 8 + 8 + 8$ or 54. Since the elements in the set add up to 91, the last term is $91 - 54$ or 37. This is the greatest possible positive integer that can be in a set with the constraints specified in the problem.

6 Gold.

- Throughout this problem we regard an $x\%$ on a test as a score of x.

- If the average of Ino's first three test scores was 93, the sum of the three scores was 93(3) or 279.

- Her average test score will increase the fastest if she scores 100s on every forthcoming test. We assume that she does this because we are looking for the minimum number of additional tests that she needs to take.

- After t more tests, the sum of her scores will be $279 + 100t$. The number of tests she will have taken is $t + 3$. Therefore, her average test score will be $\dfrac{100t + 279}{t + 3}$.

- We want this to be greater than or equal to 97, so we set up the inequality $\dfrac{100t + 279}{t + 3} \geq 97$. Multiplying both sides by $t + 3$ yields $100t + 279 \geq 97t + 291$ (we don't have to worry about flipping the sign because $t + 3$ has to be positive), and solving from here yields $t \geq 4$. Therefore, the minimum value of t is 4 tests.

7 Silver.

- The range of the sequence is equivalent to the difference between the greatest term and the least term. Since the sequence is increasing, the greatest term is the last term and the smallest term is the first term.

- We are given that the first term in the arithmetic sequence is a. The n-th term in the sequence is the last term, and it equals $a + d(n - 1)$.

- $a + d(n - 1) - a = d(n - 1)$

Chapter 2

Counting and Probability

Part 1: Basic Probability

Probability means the likelihood of something to happen. Probability is expressed as a fraction, a ratio, or a percent.

For example, if a coin is flipped, there is a $\frac{1}{2}$ or 50% chance that it lands on heads. Therefore, expect heads 50% of the time, or 1 out of every 2 tosses.

If the probability of winning a game is $\frac{1}{1000}$, expect to win 1 out of every 1000 times that you play. In one play, you are most likely not going to win.

The word probability is used interchangeably with the word chance. This is because probability is defined as the chance that something will happen.

Now that we know what probability means, we will learn how to find probabilities.

The main formula is the probability of something happening

$$= \frac{\text{the number of possible successful outcomes}}{\text{the total number of possible outcomes}}.$$

Competitive Math for Middle School: Algebra, Probability, and Number Theory
Vinod Krishnamoorthy

Copyright © 2018 Pan Stanford Publishing Pte. Ltd.
ISBN 978-981-4774-13-0 (Paperback), 978-1-315-19663-3 (eBook)
www.panstanford.com

Example 1: If a coin is tossed, what is the probability that it lands on heads?

- There are two possible outcomes of the toss of the coin—heads and tails. Therefore, the denominator of the fraction is 2.
- The only successful outcome is heads, so the numerator is 1.
- The final probability is $\frac{1}{2}$.

This formula only works if all of the outcomes are equally likely to occur. If they are not, more work must be done.

Example 2: 1000 lottery tickets are being handed out, and one of them is the winning ticket. If a person receives exactly one ticket, what is the probability that it is the winning ticket?

- What is the total number of possible outcomes? 1000 tickets are being handed out, so the person could have received any one of those 1000 tickets.
- What is the number of possible successful outcomes? Only one of the 1000 tickets is the winning ticket, so there is 1 possible successful outcome.
- Therefore, the probability is $\frac{1}{1000}$.

Example 3: 60% of the people in a building own a car. If one person is randomly selected out of the building, what is the probability that he or she owns a car?

- Since 60% or $\frac{3}{5}$ of the people in the building own a car, if a person is randomly selected out of the building, there is a $\frac{3}{5}$ chance that he or she owns a car.
- Here's why: Let us call the number of people in the room x. $\frac{3}{5}x$ of these people own a car.
- What is the total number of possible outcomes: That is x, since there are x people that can to be selected.

- What is the number of possible successful outcomes: $\frac{3}{5}x$, since there are $\frac{3}{5}x$ people that own a car.
- Therefore, the probability is $\frac{3}{5}x/x$ or $\frac{3}{5}$.

If the probability of an event is 1, it is definitely going to happen, and if the probability of an event is 0, it definitely will not occur. The probability of an event can never be greater than 1, since there can never be more possible successful outcomes than total possible outcomes.

Probability can also involve lengths and areas rather than countable outcomes. In this case, the probability would equal the successful length/area divided by the total length/area. If you understand the basic concept of probability, this should come intuitively.

Example 4: Dino awakens after a 3-year nap. What is the probability that the time is between 8:00 a.m. and 11:00 a.m.?

- Only regard the day that Dino awakens.
- There are 24 hours in a day. The window of time that represents the successful outcome is 3 hours long. Therefore, the probability is 3/24 or 1/8.

Another important formula in probability is: (the probability that an event occurs) + (the probability that it does not occur) = 1. This formula's proof uses casework, a technique you will learn later in this text.

This formula can also be interpreted as (the number of successful outcomes) + (the number of unsuccessful outcomes) = (the total number of outcomes).

Sometimes, it is easier to count the number of unsuccessful outcomes rather than the number of successful outcomes. Then, you can use this formula and the total number of outcomes to solve for the number of successful outcomes. This technique is called complimentary counting.

Example 5: There is a $\frac{1}{100}$ chance that it will rain tomorrow. Find the probability that it will not rain tomorrow.

- The probability that an event occurs plus the probability that it does not occur equals 1. Therefore, $\frac{1}{100}$ plus the desired probability equals 1.

- It follows that the answer is $\frac{99}{100}$.

Problems: Basic Probability

1 Bronze. If Vio rolls a regular cubic die, what is the probability that he will roll a six?

2 Bronze. If Vio flips a regular coin, what is the probability that it will land on heads?

3 Bronze. If Vio randomly chooses a positive integer from 1 to 100, what is the probability that he will choose neither 32 nor 74?

4 Bronze. If Vio randomly chooses a positive integer from 1 to 100, what is the probability that he will choose either 33 or 75?

5 Silver. If Vio randomly chooses a positive integer between 14 and 16 inclusive, what is the probability that it is less than 10?

6 Silver. If Vio randomly chooses a positive integer between 14 and 17 inclusive, what is the probability that it is less than 20?

7 Bronze. Vio has a bag with 7 red marbles, 5 blue marbles, and 8 green marbles. If he randomly picks a marble out of the bag, what is the probability that it is blue?

8 Bronze. A woman has as bag with 7 red marbles, 5 blue marbles, and 8 green marbles. If she randomly picks a marble out of the bag, what is the probability that it is not blue?

9 Bronze. 75% of the kids in a classroom like the color blue. If one of the students in the classroom is randomly selected, what is the probability that he or she likes blue?

10 Bronze. $\frac{1}{5}$ of the marbles in a bag are red. If you randomly select one marble out of the bag, what is the probability that it will be red?

11 Silver. A girl picks three numbers from 1 to 100. A boy then picks four more numbers from 1 to 100. All of the numbers picked are different. What is the probability that the girl picked the highest number?

12 Bronze. An element is randomly chosen out of the set $\{0,2,2,4,5,7\}$. What is the probability that the chosen element is 5?

13 Bronze. The area of a dartboard is 12 square units. The area of the bulls-eye region on the dartboard is 2 square units. If a dart is randomly thrown somewhere on the dartboard, what is the probability that it hits the bulls-eye region?

14 Silver. A ball is rolled along a straight carpet that is six feet long. It must stop somewhere along the carpet, and it is equally likely to stop at every place. What is the probability that the ball stops between 3 feet and 5 feet from its starting position?

15 Bronze. Dino awakens after a 3-year nap. What is the probability that the time is *not* between 8:00 a.m. and 11:00 a.m.?

Part 2: Basic Counting

Sometimes, counting outcomes one by one takes forever to complete. Here are some counting principles that will save you endless amounts of time.

When there are *n* ways to do something and *m* ways to do another thing, the number of ways to do <u>both *n* and *m*</u> are $n \times m$. However, the number of ways to do either one thing *or* the other is $n + m$.

Example 1: Nid is going to an ice cream shop. He is going to pick one flavor of ice cream and one topping. If there are 3 different flavors of ice cream and 5 different toppings offered by the ice cream shop, how many different ice cream- topping combinations can he make?

- There are 3 ways to pick a flavor of ice cream, and 5 ways to pick a topping. Therefore, there are 3 × 5 or 15 ways for Nid to pick a flavor of ice cream <u>and</u> a topping.

Example 2: A child is in an ice cream shop. There are five premium toppings, four normal toppings, and three economic toppings. The child can choose one topping for his ice cream. How many ways are there for him to do this?

- The child can pick a premium topping, a normal topping, or an economic topping. Therefore, there are 5 + 4 + 3 = 12 ways for him to do this.

Here is another important counting principle: If there are *m* things <u>for each of</u> *n* things, then the total number of things is *n* × *m*.

For example, if there are 100 words per document, and there are 10 documents, then there are 100 × 10 = 1000 words in total. This principle closely relates to the "and" principle we just discussed.

Problems: Basic Counting

1 Silver. Kann is at a turning point in a road. He can turn either left or right. If he turns left, there are three pathways that he can take, and for each of these pathways there are five sub-pathways that all lead to his destination. If he turns right, there are 3 pathways that all go directly to his destination? How many total paths can Kann take to his destination?

2 Bronze. There are two rooms next to each other in an office. The first room has 50 boxes of dolls. Each box of dolls has 4 dolls inside. The second room has 34 boxes of dolls. Each of those boxes has 2 dolls inside. How many total dolls are there in both the rooms?

3 Silver. An outfit consists of one shirt, one pair of shorts, and one pair of shoes. A boy has 5 shirts, 10 pairs of shorts, and 2 pairs of shoes. How many distinct outfits can he make?

4 Gold. How many numbers are there with tens digit 5 between 10 and 999?

5 Silver. There are two desks in a room. On top of one desk are two English books, three Spanish books, and six French books. On top of the other desk are three English books and three French books. If a woman comes in and picks one book at random from each of the desks, what is the probability that they are both English?

6 Bronze. There are 10 grams of sugar in one liter of a certain drink. How many total grams of sugar are in 20 liters of the drink?

Part 3: Multiple Events

We already know how to find the probability of one event. But what if we are asked to find the joint probability of multiple events?

To find the probability that two events both occur, multiply their individual probabilities. This collapses the two events into one event with a specific probability. A simple type of combination of events is independent events. If the outcomes of two events do not affect each other, the events are independent.

Example 1: Jim flips a coin twice. What is the probability that he receives two heads?

- Since the probability of receiving heads on each flip is $\frac{1}{2}$, the probability of receiving heads on both flips is $\frac{1}{2} \times \frac{1}{2}$ or $\frac{1}{4}$.

Example 2: Today, there is a $\frac{1}{5}$ chance that it will rain and a $\frac{1}{3}$ chance that the pool will be closed for maintenance. What is the probability that at least one of the two happens?

- The probability that at least one of these things happens equals the probability that neither happens subtracted from 1.

- The probability that it will not rain is $1 - \dfrac{1}{5}$ or $\dfrac{4}{5}$, and the probability that the pool will not be closed for maintenance is $\dfrac{2}{3}$.

- Their combined probability is $\dfrac{4}{5} \times \dfrac{2}{3}$ or $\dfrac{8}{15}$; $1 - \dfrac{8}{15} = \dfrac{7}{15}$.

Here is a justification of this principle using counting techniques.

- Define event A with b possible outcomes and a successful outcomes. Also define event B with d possible outcomes and c successful outcomes.

- By the definition of probability, the probability of event A is $\dfrac{a}{b}$ and the probability of event B is $\dfrac{c}{d}$.

- Now we will find the probability that both of these events occur. One possible outcome must occur out of event A, and one must occur out of event B. The number of ways to choose one of the b possible outcomes of event A <u>and</u> one of the d possible outcomes of event B is $b \times d$.

- To count the number of ways that successful outcomes will occur in both events, we choose one of the a successful outcomes of event A and one of the c successful outcomes of event B. The number of ways to do this is $a \times c$.

- Therefore, the probability that both event A and event B will occur is $\dfrac{ac}{bd}$, which equals $\left(\dfrac{a}{b}\right)\left(\dfrac{c}{d}\right)$, the product of the events' respective probabilities.

Problems: Multiple Events

1 Bronze. If Ivonde rolls a regular dice and tosses a quarter, what is the probability that she will roll an even number and toss a heads?

2 Silver. 4 stacks of 40 cards are laid out on a table. 10 of the cards in each stack are special cards. Ivonde picks one card from each stack. She needs a special card in the first, second, and third stacks and a card that is not special from the fourth stack to win a prize. What is the probability that she wins?

3 Bronze. Ivonde flips four coins. What is the probability that all of them land on heads?

4 Bronze. There are 8 purple lamps, 4 blue lamps, and 2 yellow lamps in a shop. If two lamps are randomly chosen from this set with replacement, what is the probability that the first is blue and the second is purple?

5 Silver. A man has two buttons right in front of him, one red and one blue. What is the probability that the third time he presses any button, it is the blue one?

6 Silver. There are five different flavors of ice cream in an ice cream shop. Ino and Vod each go to the shop at different times and buy one flavor of ice cream. What is the probability that they buy the same flavor?

7 Silver. There are 4 rows of 3 cups on a table. Pebbles are underneath two cups in each row. Vind picks one cup from each row, and then the arrangement is reset. A while later, Div comes in and does the same. What is the probability that all eight cups they pick will have pebbles underneath them?

8 Bronze. A gorilla rolls 4 standard die. What is the probability that all of them land on an even number?

9 Gold. The probability of a certain event occurring is p. The probability of another event independent of p occurring is q. What is the probability that p will occur but q will not? Express your answer in terms of p and q.

Part 4: Orderings

In mathematics, an exclamation mark does not add excitement to anything. Instead, it is the factorial operation. To perform this operation, find the product of the number before the exclamation point and every positive integer less than it. For example, $5! = 5 \times 4 \times 3 \times 2 \times 1 = 120$, and $10! = 10 \times 9 \times 8 \times 7 \times 6 \times 5 \times 4 \times 3 \times 2 \times 1$. This operation is very common in counting and probability.

Example 1: There are 5 cards labeled A, B, C, D, and E, respectively. How many ways are there to order the five cards?

- We start by finding the number of ways to place the card labeled A in an arrangement of the cards. The A card can be first, second third, fourth, or fifth. Therefore, there are 5 ways to place the A card.

- After we place the A card, we must consider the B card. After placing the A card first, second, third, fourth, or fifth, there are 4 options left. The B card can be any one of the four, so there are 4 ways to place the B card.

- We see there are $5 \times 4 = 20$ ways to place both the A and the B cards within the group of 5 cards.

- Similarly, there are 3 ways to place the C card, 2 ways to place the D card, and 1 way to place the E card.

- Therefore, there are $5 \times 4 \times 3 \times 2 \times 1$ ways to order the five cards. This is the same as 5 factorial!

From this problem, we see that the number of ways to order n distinct things is $n!$.

However, what if multiple things are identical?

Example 2: How many ways are there to order the letters A, B, B, C, and D.

- If we try to order these using our first method, we will get numerous repeats. For example, A, C, B, D, B is the same thing as A, C, B, D, B.

- But how many repeats are there? For every different ordering of the full set, there are 2! different ways to order the two B's within it. Swapping the B's with one another results in exactly the same full sequence.
- Therefore, we must divide 5! by 2! to obtain the final answer of 60.

The number of ways to order n things with m things repeated is $n!/m!$.

However, there can be multiple sets of repeated things, such as in the set {A,A,A,B,B,B,C}. If this occurs, divide by multiple $m!$ terms. In this case, the number of possible orders is 7!/3!/3!, since both A and B are repeated three times.

Problems: Orderings

1 Bronze. How many different four-digit positive integers can be made out of the digits 1, 2, 3, and 4?

2 Bronze. There are five differently colored jellybeans in a bowl: one red, one blue, one yellow, one green, and one purple. Ino plays a game in which she draws all the jellybeans out, one by one. She wins if the order they are selected in is red, green, blue, yellow, purple or red, yellow, purple, blue, green. In one round, what is the probability that she will win?

3 Silver. Find the value of $\dfrac{29!}{28!}$.

4 Silver. How many different seven-digit positive integers can be made with the digits 1, 2, 2, 2, 3, 4, and 5?

5 Bronze. There are four different chairs and four different cushions in a room. In how many ways can Von put one cushion on each chair?

6 Silver. There are three rows of small boxes. The first row has 3 boxes, the second row has 4, and the third row has 5. One red coin, one blue coin, and one yellow coin are going to be placed in one of the boxes. Only one coin can be in each row. In how many ways can this happen?

7 Silver. There are six chairs and six cushions in a room. All of the chairs are distinct. Three of the cushions are identical, but the other cushions are distinct. In how many non-identical ways can Von put one cushion on each chair?

8 Platinum. The digits of the positive integer 1234 can be reordered to form many other positive integers. Some examples are 1243, 3124, and 4312. What is the sum of all of these positive integers, including 1234?

Part 5: Dependent Events

Dependent events are the opposite of independent events. In a system of dependent events, the outcomes of the events affect each other. There are many kinds of problems involving dependent events; we will go through the most common ones.

Casework is the process of splitting a complex event into different cases that are manageable, finding the probability of each case, and *adding* these probabilities to obtain the probability of the entire event.

Example 1: Node plays a game in which he flips two coins. He wins if he receives either two heads or two tails. What is the probability that Node will win?

- We split this problem into two cases.

- If the first coin yields heads, the second coin must yield heads as well. This is a simple system of independent events with probability $\frac{1}{4}$.

- The other case is if the first coin yields tails. For Node to win, the second coin has to yield tails as well. The probability of this happening is also $\frac{1}{4}$.

- We add the probabilities of the two cases, obtaining $\frac{1}{2}$ as the answer.

When doing casework, why do we add the probabilities of the cases at the end? Here is a mathematical justification:

- Let us define event A with d possible outcomes.
- We split event A into two exclusive cases, meaning that if one case occurs the other case cannot. Let us call the number of successful outcomes in the first case a and the number of successful outcomes in the second case b.
- The probability of the first case is $\frac{a}{d}$, and the probability of the second case is $\frac{b}{d}$.
- What is the total number of successful outcomes of event A? The number of ways to choose one of the a successful outcomes of the first case or the b successful outcomes of the second case is $a + b$.
- Therefore, the probability of the entire event is $\frac{a + b}{d}$.
- This equals $\frac{a}{d} + \frac{b}{d}$, which is the sum of the probabilities of the two individual cases.

To use casework in a problem, you must make sure that no two cases can occur simultaneously. Otherwise, the cases will not be exclusive, and you will have to use another method to find the final probability.

Example 2: Viena is searching for a certain document in a filing cabinet with three drawers. There is a $\frac{1}{5}$ chance that the document is in the top drawer, a $\frac{1}{20}$ chance that it is in the middle drawer, and a $\frac{1}{10}$ chance that it is in the bottom drawer. What is the probability that the document is in this particular filing cabinet?

- We treat Viena finding the document in each of the three drawers as individual cases of her finding the document in the whole filing cabinet.
- The document cannot be in more than one drawer at once. Therefore, the three cases specified in the problem are exclusive.

- To find the probability of the entire event, we add the probabilities of its cases. Doing this yields $\dfrac{7}{20}$ as the final probability.

In addition to other things, casework is useful for finding probabilities where the possible outcomes are not equally likely.

A red flag in casework arises if your answer, after adding the probabilities of the cases, is over 1, since the probability of any event cannot be larger than 1.

We can also use the principles of casework in counting.

Example 3: Ino is trying to make a three-digit positive integer. If the hundreds digit is odd, then the two other digits have to be the same. If the hundreds digit is even, then the two other digits have to be different. How many different three-digit positive integers can she make under these restrictions?

- To solve this problem, split the problem into two different cases; the first when the hundreds digit is odd, and the second when the hundreds digit is even.

- There are 5 possible hundreds digits in the first case: 1, 3, 5, 7, and 9. After this, there are 10 possibilities for the remaining digits: 00, 11, 22, etc. Therefore, there are $10 \times 5 = 50$ successful integers in the first case.

- In the second case, there are 4 possible hundreds digits: 2, 4, 6, and 8.

- For each hundreds digit, there are 10 possible tens digits, and then after that is chosen, there are 9 possibilities ones digits (since it cannot be the same one as the tens digit).

- Therefore, the total number of possibilities in this case is $4 \times 10 \times 9 = 360$.

- Adding the probabilities of the two cases yields $50 + 360$ or 410, so Ino can make 410 positive integers.

Another form of dependent events problems involves choosing without replacement.

Let us say you are picking 3 marbles from a bag of marbles. If you use replacement, then you put the marble you pick back into the bag each time.

Choosing multiple times with replacement are independent events, because the pool of choices resets each time you choose. But choosing without replacement forms dependent events.

Let us say that you have a bag of 20 different marbles, and you are picking marbles out of it without replacement. There are 20 marbles to pick for the first marble, but after picking the first marble, there are only 19 left. After picking the second, there are only 18 left. Understanding this principle is very important.

Example 4: Node has a bag of 20 red marbles and 10 blue marbles. He picks three of these without replacement. What is the probability that they are all red?

- The probability that the first marble is red is $\dfrac{20}{30}$.
- When choosing the second marble, one red marble is gone. Therefore, the probability that the second marble he picks is red is $\dfrac{19}{29}$, and likewise the probability that the third marble he chooses is red is $\dfrac{18}{28}$.
- Multiplying all of these, we find that the final probability is $\dfrac{57}{203}$.

Problems: Dependent Events

1 Gold. Vode is playing a game with a dice. He wins if he rolls a three. He has three rolls to do this, and as soon as he wins he will stop. What is the probability that he will win?

2 Silver. There are 8 blue marbles and 7 red marbles in a jar. If two marbles are randomly selected from the jar without replacement, what is the probability that both marbles are red?

3 Silver. There are four purple lamps, three blue lamps, two red Frisbees, and five yellow Frisbees in a shop. If two lamps and two

Frisbees are randomly selected without replacement, what is the probability that both of the lamps are purple and both Frisbees are yellow?

4 Bronze. Vode is playing a game in which he flips a coin and rolls a standard die. He wins if he gets tails and a 3 or heads and a 6. What is the probability that Vode will win?

5 Gold. Vode goes to the gym to lift weights. The probability that his arms will be sore on the first day is $\frac{2}{3}$. If his arms become sore, then he will stay home the next day, but if his arms are not sore, then he will go to the gym the next day fully refreshed. If he goes to the gym for a second time, the probability that his arms will become sore there is $\frac{2}{9}$. What is the probability that Jim's arms become sore on the second day?

6 Gold. 80% of the kids in a classroom like the color blue, and 25% of the kids in the classroom like both blue and red. Given that liking blue and liking red are completely independent of one another, what is the probability that a kid who likes blue also likes red?

7 Gold. Below is a grid of 16 unit squares. If two unit squares are chosen at random, what is the probability that they share a vertex?

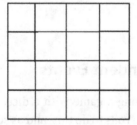

8 Silver. There are 4 friends in a juice shop. Each is randomly given one of three possible drinks. What is the probability that all four of the friends get the same drink?

9 Gold. There are 4 purple and 5 blue jellybeans in a bowl. Ino picks 9 jellybeans out of the bowl, one at a time. What is the

probability that she will pick an alternating sequence of purple and blue jellybeans?

10 Platinum. There are 9 cards in a bag. Exactly one of these cards is yellow. A girl and a boy are playing a game where they take turns picking a card out of the bag and then replacing it. The first one that picks a yellow card wins, and the game is over. If the girl goes first, what is the probability that she will win?

11 Gold. A person is going to pick two integers between 1 and 100 inclusive (inclusive means that 1 and 100 are both valid picks). They do not necessarily have to be distinct. What is the probability that the product of the numbers is even?

12 Silver. Vino, Inov, Oniv, Niov, and Dio each take a seat in a row of five seats. If they choose their seats randomly, what is the probability that Oniv sits next to Vio.

13 Silver. A special coin has a $\frac{2}{3}$ chance of landing on heads and a $\frac{1}{3}$ chance of landing on tails. What is the probability that in two tosses, one heads and one tails in either order will be obtained?

14 Gold. Node and Ino are each picking whole numbers. Ino randomly picks two distinct numbers between 1 and 50 inclusive, and Node randomly picks another number between 26 and 50 inclusive. Node's number cannot be the same as either of Ino's. What is the probability that Node picks the highest number?

15 Gold. If the probability that event A occurs is p, and the probability that independent event B occurs is q, find the probability that either event A or event B, but not both, occurs in terms of p and q.

16 Silver. The probability that the janitor forgets to unlock the gym on a Friday is $\frac{1}{500}$, and the probability that the volleyball team has planned to practice on a Friday is $\frac{1}{3}$. The basketball team can practice only if neither of these things happens. On a given Friday, what is the probability that the basketball team can practice?

17 Silver. There are 4 yellow candies and 10 green candies in a jar. If two candies are randomly picked out of the jar without replacement, what is the probability that at least one of them will be yellow?

Part 6: Subsets

Let us say that I have a set of toys. A subset of this set is a set of toys taken out of the bigger set. The subset can have any number of toys less than or equal to the number of toys in the original set.

Sets can be of anything; toys, burgers, numbers, variables, etc. An element is an item in a set. Sets of numbers or variables are written as elements separated by commas. The whole set is enclosed by { }.

Consider the set {1,2,3,4,5}. Possible subsets include {1}, {2,4}, and {1,2,3,4,5}.

Example 1: Find the number of subsets of {1,2,3,4,5}.

- To calculate the number of subsets of the bigger set, we examine each individual element.

- For 1, there are two possibilities. Either 1 is in the set, or it is not. The same goes for 2, 3, 4, and 5.

- Therefore, there are $2 \times 2 \times 2 \times 2 \times 2$ or 32 subsets of this set in total. This includes a common subset, the empty set { }, that has no elements.

A formula for finding the total number of subsets of a set with n distinct elements is 2^n. But what if there are repeated elements?

Example 2: Find the number of subsets of {1,2,2,4,5}.

- If we use our first formula, we count many identical sets more than once.

- Let us go back analyzing each element. There are 4 different elements in this set: 1, 2, 4, and 5. There are two possibilities each for 1, 4, and 5 (as shown earlier).

- However, there are more than two possibilities for 2, since there are two 2's. You can have no 2's, one 2, or both 2's in any subset.
- Therefore, the total number of subsets is $2 \times 3 \times 2 \times 2 = 24$.

Problems: Subsets

1 Bronze. Find the number of subsets of the set $\{1,2,3,4\}$?

2 Bronze. Find the number of non-empty subsets of the set $\{3,9,90,12,43\}$?

3 Silver. Find the number of subsets of the set $\{1,1,2,2,3,3,3,4,5\}$?

4 Silver. Onid is at a shop with two identical fans, two identical tables, and one chair. How many different combinations of items can he purchase?

5 Gold. There are 4 different spices and 3 different sweeteners that Suu'haak can use in his soup. His soup must contain at least one spice and one sweetener. How many possible combinations of spices and sweeteners can he add to his soup?

Part 7: Organized Counting

Counting outcomes is an essential part of probability. You may think it is simple, but as problems get more complex, counting requires considerable skill.

Example 1: Vio rolls two standard six-sided dice. How many ways are there for him to get a sum of 6, 7, or 8?

- The conditions are satisfied if he rolls a 1 and a 5, if he rolls a 3 and a 4, and also if he rolls a 4 and a 3. However, this method of randomly listing is not the best way to do this problem, since it will be easy to miss a few possibilities.
- Let us try to create an organized list. We start with the possibilities that add up to 6.

- The pairs that add up to 6 are 1 and 5, 2 and 4, 3 and 3, 4 and 2, and 5 and 1.

- Now the pairs that add up to 7: 1 and 6, 2 and 5, 3 and 4, 4 and 3, 5 and 2, and 6 and 1.

- Lastly, we find the pairs that add up to 8: 2 and 6, 3 and 5, 4 and 4, 5 and 3, and 6 and 2.

Notice that we put all the pairs we counted into a specific order so that we would not miss any. Counting all the pairs up, we find that there are 16 ways for Vio to get a sum of 6, 7, or 8.

Example 2: List all non-empty subsets of the set $\{1,3,4\}$.

- In order to be organized, we start with the one-element subsets, then go to the two-element subsets, and finish with the three-element subsets.

- The one-element subsets are $\{1\}$, $\{3\}$, and $\{4\}$ in numerical order.

- The two-element subsets are $\{1,3\}$, $\{1,4\}$, and $\{3,4\}$ in numerical order.

- The only three-element subset is $\{1,3,4\}$.

This method of creating organized lists can be used in many scenarios. Always try and find a way to organize your counting so that you neither over-count nor under-count.

Problems: Organized Counting

1 Silver. If I write all of the whole numbers from 1 to 99 on a piece of paper, how many times will I write the digit 9?

2 Bronze. List all non-empty subsets of the set $\{w, x, y, z\}$.

3 Gold. How many three-digit positive integers exist where the digits add up to 12?

4 Bronze. If two dice are rolled, what is the probability that the sum of the numbers on the top faces will be 7?

5 Silver. There is a big box of bags of marbles. Some of the bags contain 3 marbles, some of them contain 5, and some of them contain 7. There are plenty of each. Vode will pick three bags from the box, and then dump the contents of the bags onto his desk. How many possible numbers of marbles can be on his desk?

6 Silver. A palindrome is a number that is the same read forward and backward. For example, 136631 and 262 are palindromes. How many four-digit palindromes contain no odd digits?

7 Silver. Vode is playing basketball. He can make either 3-point shots or 2-point shots in the game, nothing else. How many ways are there for him to score 36 points in the game?

8 Silver. How many ways are there to get a product of 100 using two positive integers from 1 to 75 if their order does not matter?

9 Silver. How many sets of 2 positive integers add up to 50?

10 Gold. How many ways are there to make $0.85 with nickels, dimes, and quarters?

11 Platinum. A set made up of 7 positive integers has a unique mode of 5 and a mean of 8. What is the greatest possible number that can be in this set?

12 Silver. There are 5 blue candies, 4 red candies, 9 orange candies, and 13 yellow candies in a jar. What is the least number of candies that a person has to pick out of the jar to guarantee that he or she has at least 3 yellow candies?

Part 8: Permutations and Combinations

Let us say that a certain number of items are being chosen out of a larger group. This is where combinations come into play. There are two main types of combinations: those where order matters and those where it does not.

Example 1: You have a group of 7 items, and you are going to choose four of them, one by one. How many ordered combinations can you receive?

- For the first choice you have 7 options, for the second you have 6, for the third you have 5 and for the fourth you have 4. This makes a total of $7 \times 6 \times 5 \times 4$ or 840 ordered combinations.

When m things are being selected from a larger set of n things and the order in which they are selected matters, the number of possible combinations is $\dfrac{n!}{(n-m)!}$.

We would use this formula in the previous problem by setting $n = 7$ and $m = 4$. $\dfrac{7!}{(7-4)!}$ or $\dfrac{7!}{3!}$ equals $7 \times 6 \times 5 \times 4$ due to 3, 2, and 1 canceling out in both the numerator and denominator. Combinations where order matters are called permutations. The formula $\dfrac{n!}{(n-m)!}$ is represented as $^{n}P_{m}$, $P_{n,m}$, or "n permute m".

Example 2: There are 10 differently colored cubes in a box. Oino makes a sequence of cubes by choosing three cubes out of the box, one at a time. How many different sequences are possible?

- The problem calls for the number of ways that three things can be selected out of a set of 10 where order does matter. The answer is $^{10}P_{3}$, which equals $\dfrac{10!}{7!} = 10 \times 9 \times 8 = 720$.

However, what if the order in which the items are chosen does not matter?

If we take $7 \times 6 \times 5 \times 4$ as in the previous example, we overcount. Since there are 4! or 24 orderings of each unordered combination of four items we choose, the product $7 \times 6 \times 5 \times 4$ counts each combination we want 24 times.

Therefore, we divide 840 by 24 to obtain our answer, 35.

To choose m things out of a set of n things where the order in which they are chosen does not matter, use the formula $\dfrac{n!}{m!(n-m)!}$. It is just like the previous formula except with an additional $m!$ in the denominator.

This formula has a name as well. It is called the "choose" or combination function, and is represented with a C ($^{n}C_{m}$ or $C_{n,m}$) or with the two numbers stacked on top of one another in parentheses as in $\binom{n}{m}$.

n choose m can also be thought of as multiplying the first m terms of $n!$ and then dividing that by $m!$. This equals our original expression due to the fact that the remaining terms in the original expression end up canceling each other out. Using this logic, 8 choose 3 would equal $\dfrac{8 \times 7 \times 6}{3!}$, and 11 choose 5 would equal $\dfrac{11 \times 10 \times 9 \times 8 \times 7}{5!}$. This way of thinking about the choose function is faster than individually calculating each part of the entire expansion.

Example 3: Find the number of three element subsets of a set with 9 distinct elements.

- In a subset, the order of the elements does not matter. The number of three element subsets of a set with 9 distinct elements equals the number of ways to choose three things out of nine distinct things, or $\binom{9}{3}$. This equals $\dfrac{9 \times 8 \times 7}{3!} = 84$.

Choosing can be used in many different ways.

Example 4: In how many ways can we arrange three objects into five slots if exactly two of the objects are identical?

- To solve this problem, we choose two slots to put the identical objects, and then one of the remaining slots to place the other object.

- There are 5 choose 2 ways to choose slots for the two identical objects. Since the object are identical, the order in which they go in the two designated slots does not matter. Therefore, every distinct arrangement of the two identical objects is represented by the number of ways there are to choose 2 slots out of the 5. $\binom{5}{2} = \dfrac{5 \times 4}{2!} = 10$.

- Now we worry about the third object that is not identical to the other two. *For each* of the 10 arrangements of the two identical objects, there are 3 slots left over that the third can occupy.

- Therefore, the total number of possible arrangements is 10×3 or 30.

Example 5: The probability that a cereal box has a gold star inside it is $\frac{2}{7}$. If a person buys four cereal boxes, what is the probability that he or she receives exactly two gold stars?

- Let us label the boxes as Box 1, Box 2, Box 3, and Box 4.
- The probability of obtaining gold stars from Box 1 and Box 2 but not Box 3 and Box 4 is $\frac{2}{7} \times \frac{2}{7} \times \frac{5}{7} \times \frac{5}{7}$, or $\frac{100}{2401}$.
- However, this is not the only case that satisfies our requirements. What about obtaining gold stars from Box 1 and Box 3, or obtaining gold stars from Box 3 and Box 4?
- There are $\binom{4}{2}$ or 6 ways to choose two of our boxes to have a gold star in them. Therefore, there are 6 total cases, each with a probability of $\frac{100}{2401}$, that satisfy the requirements of this problem.
- $\frac{100}{2401} \times 6$, or $\frac{600}{2401}$, is our answer.

In counting problems, it is extremely important to regard whether order matters or not.

When calculating probability, if the problem implies that order does not matter, but you regard order in both the numerator and denominator, the answer should come out to be the same. Reversing this principle can often cause problems, so only use it as stated above.

A very well-known theorem called the Binomial Theorem relates Pascal's Triangle and expanding binomials to the choose function.

This theorem states that $(a + b)^n = \binom{n}{0} \times a^n b^0 + \binom{n}{1} \times a^{n-1} b^1 + \binom{n}{2} \times a^{n-2} b^2 + \ldots + \binom{n}{n-1} \times a^1 b^{n-1} + \binom{n}{n} \times a^0 b^n$.

Compare this formula to Pascal's Triangle to see how they really are the same thing.

Also note that 0! is mathematically defined to be equal to 1, and anything choose 0 equals 1. This makes sense, as there is 1 way to order 0 things.

Problems: Permutations and Combinations

1 Bronze. Find the value of $^9P_3$.

2 Bronze. Find the value of $^6C_4$.

3 Silver. Find the value of 95 choose 94.

4 Bronze. There are 20 different garden gnomes in a store. A boy is going to purchase two of them. How many different pairs can he purchase?

5 Silver. Inode has 12 bricks, each painted with a different pattern. She chooses 4 of these bricks, one at a time, and stacks the bricks in the order that she chose them. How many different stacks can she obtain?

6 Silver. Vio is playing a game in which she randomly picks two index cards out of a jar of 22. Five of the 22 cards in the jar have blue dots on them. For Vio to win the game, one or more of the cards she picks must have a blue dot. In one play, what is the probability that Vio will win?

7 Silver. Find the number of three-element subsets of a 13-element set with distinct elements.

8 Silver. A group of people has 25 hats in front of them. They first choose a subset of 10 hats to take back to their car. Then, they choose 5 of those 10 hats to return back to the store. Lastly, they choose 3 of those 5 hats to keep in their homes. How many sets of 3 hats can be chosen?

9 Gold. If someone flips 11 coins, what is the probability that they will receive 8 heads and 3 tails (in any order)?

10 Gold. A girl has three markers of different colors and one black pen. She is going to color a grid of six white squares. Three of the squares are going to be colored with the markers, and the remaining three will be either colored black or left white. The black pen can be used multiple times, but each marker can only be used to color one square. How many different ways are there to color the grid?

11 Silver. A person randomly selects 3 of the first 15 positive integers. What is the probability that the person selects neither 8 nor 13?

12 Gold. In a specially weighted coin, the probability of landing on heads is $\frac{1}{3}$ and the probability of landing on tails is $\frac{2}{3}$. If this coin is flipped 8 times, what is the probability that the coin will land on tails exactly four times?

13 Gold. At a gathering, everyone exchanges phone numbers with everyone else exactly once. If numbers were exchanged 153 times, how many people were at the gathering?

14 Gold. Explain why it is true that $\binom{n}{0} + \binom{n}{1} + \binom{n}{2} + \binom{n}{3} + \ldots + \binom{n}{n-1} + \binom{n}{n} = 2^n$.

15 Silver. What are the first three terms of $(a + 3b)^{11}$ when in standard order?

Part 9: Expected Value

What is an "expected value"? Expected value is a concept involving probability. It is defined as the expected or average amount of something to be obtained in a probability-based event.

How is it calculated? First, find all of the different amounts that are possible to be obtained. Next, multiply each of those amounts by the respective probability of obtaining it. Add all of the results and you have your expected value. This will make more sense with an example.

Example 1: Two bags each contain three cards numbered 1 through 3. If one card is randomly drawn from each bag, what is the expected sum of the cards' numbers?

- What are the possible sums? The smallest possible sum is 2 (1 and 1), and the greatest possible sum is 6 (3 and 3). Every single positive integer in between 2 and 6 can also be obtained as a sum.

- Now we calculate the probabilities of obtaining each of these sums. There are 3×3 or 9 total ways to pick one card from each bag (we differentiate the picks from the two bags, for example, we consider 1 from the first bag and 2 from the second bag different from 1 from the second bag and 2 from the first bag).

- There is only one card combination that yields a sum of 2: 1 and 1. Therefore, the probability of obtaining a sum of 2 is $\frac{1}{9}$.

- There are 2 ways to obtain a sum of 3: 1 from the first bag and 2 from the second, and 2 from the first bag and 1 from the second. The probability of obtaining a sum of 3 is therefore $\frac{2}{9}$.

- Going through this process for the other 3 possible sums, we find that the probability of obtaining a sum of 4 is $\frac{3}{9}$ or $\frac{1}{3}$, the probability of obtaining a sum of 5 is $\frac{2}{9}$, and the probability of obtaining a sum of 6 is $\frac{1}{9}$.

- Therefore, the expected sum is $2 \times \frac{1}{9} + 3 \times \frac{2}{9} + 4 \times \frac{1}{3} + 5 \times \frac{2}{9} + 6 \times \frac{1}{9}$. This equals 4.

Note that the expected value of a scenario may not be attainable. For example, if an integer is randomly chosen between 0 and 5 inclusive, the expected value of the result is 2.5, although 2.5 is not one of the possible choices.

Recall the definition of probability: If the probability of an event is $\frac{x}{y}$, it is expected to occur x out of every y times.

Example 2: If there is a $\frac{1}{5}$ chance that it will rain on any given day, it is expected to rain on one in every five days. How many rainy days would one expect in a 15-day period?

- By the definition of probability, it is expected to rain one in every five days.

- We set up the proportional equation $\dfrac{1 \text{ rainy day}}{5 \text{ total days}} = \dfrac{x \text{ rainy days}}{15 \text{ total days}}$. The units match up in both the numerators and denominators, so they can be disregarded when solving for x.

- Solving, we find that $x = 3$ rainy days. This is a different but more straightforward type of expected value problem.

Problems: Expected Value

1 Gold. 3 dice are rolled, and the numbers that the dice land on are recorded. What is the expected number of occurrences of 1?

2 Gold. There are 12 quarters, 10 dimes, and 5 pennies in a bag. If Viod randomly picks one coin out of the bag, what is the expected amount of money that he will receive expressed to the nearest cent?

3 Silver. Eight players are playing in a single-elimination tournament. They all have equal chances of winning each match. The winner is going to receive $24, and the others will receive nothing. What is the expected amount of money that a given player will receive?

4 Silver. Every time Dinov plays a certain card game, the probability he wins is $\dfrac{1}{3}$. If he plays 24 games, how many wins should he expect?

Part 10: Weighted Average

Probability introduces another way of calculating averages. The way to calculate the *weighted average* of a set is to multiply each element by its *weight* and add all of the results.

An element's weight can be its fraction of occurrence/duration or the probability of obtaining the element.

Example 1: Calculate the average of 3, 4, and 5.

The way to calculate the weighted average is to multiply each number by its fraction of occurrence and then find the sum of the results.

The fraction of occurrence of 3 is $\frac{1}{3}$, since there are three total numbers and one of them is 3. Also, the probability of choosing 3 out of the three numbers is $\frac{1}{3}$. The same is true for 4 and 5. $\left(3 \times \frac{1}{3}\right) + \left(4 \times \frac{1}{3}\right) + \left(5 \times \frac{1}{3}\right) = 4$, our answer.

Example 2: A car travels at 45 miles per hour for 1 hour, 60 miles per hour for 2 hours, and 10 miles per hour for half an hour. Find its average speed.

The car travels for a total of 3 and a half hours. The car travels at 45 miles per hour for $\frac{1}{3.5}$ or $\frac{2}{7}$ of the total time. Similarly, the car travels at 60 miles per hour for $\frac{4}{7}$ of the total time, and at 10 miles per hour for $\frac{1}{7}$ of the total time.

We multiply each speed by its respective fraction of duration and add the results to find the car's average speed.

$$\left(45 \times \frac{2}{7}\right) + \left(60 \times \frac{4}{7}\right) + \left(10 \times \frac{1}{7}\right) = \frac{340}{7} \text{ miles per hour.}$$

Problems: Weighted Average

1 Silver. I scored a 93% on a 200-point test, an 89% on a 100-point test, and a 78% on another 100-point test. What is the weighted average of these three test scores?

2 Silver. An antique shop holds many vases. Half of the vases cost $1000 each, one-fourth of the vases cost $2000 each, one-fifth of the vases cost $3000 each, and the remaining one-twentieth of the vases cost $500 each. What is the average price of all of the vases in the shop.

3 Platinum. Now that you know about weighted average, explain why no matter how much a car varies in speed, its average speed multiplied by the time that it travels equals the total distance that it travels.

Part 11: Generality

Note: This section is more difficult than the rest of the text.

In counting, generality is an important concept to understand. If a method of counting possesses generality, it applies to all possible cases and does not undercount or overcount. Sometimes, we can reduce the number of cases that we count *without loss of generality*, meaning that all cases are still accounted for. Applying generality can save lots of time and prevent over-counting.

Example 1: How many ways are there to seat two different people in a circular table with 6 seats? Rotations of the same arrangement are not considered distinct.

- Let us label the seats 1–6. If we put the first person in seat 1 and the second person two seats away in seat 3, this arrangement is exactly the same as putting the first person in seat 2 and the second in seat 4, the first person in seat 3 and the second in seat 5, and so forth.

- If we assign the first person to seat 1 so that he/she does not move from that seat, the generality of our counting will not be compromised, as every arrangement of the two people where the first person is in seat 1 can be rotated so that the first person is in a different seat of our choosing.

- Therefore, there are 5 ways to seat the two people, as we can put the second person in any of the other 5 seats.

- After doing this, we must check to make sure that none of the 5 arrangements can be rotated to form another one of the 5 other arrangements that we counted.

Example 2: How many sets of three positive integers have a product of 50?

- Let us call the three numbers a, b, and c. Without loss of generality, let us make $a \leq b \leq c$. We can do this because the order of the three numbers does not matter, so every arrangement of a, b, and c where it is not true that $a \leq b \leq c$ is identical to an arrangement where it is.

- Knowing this, a has to be either 1 or 2, because 3 multiplied by anything cannot form 50 and 4 is too high for it to be true that $a \leq b \leq c$.

- From here, it is easy to count systematically and see that the answer is 4 sets.

Example 3: How many sets of three integers have a product of 50?

- Positive integers are always positive, but integers in general can be negative. Three positive integers multiply to form a positive integer, and so do one positive integer and two negative integers.

- We already know that there are 4 sets of three positive integers that multiply to 50. How many sets of one positive integer and two negative integers multiply to 50?

- Every set of two negative integers and one positive integer that multiplies to 50 can be formed by negating two elements of a set of three positive integers (from Example 2) that multiplies to 50. Since some of these sets contains a repeated element, we cannot generalize a result and we must count individually.

- We can reconfigure the set $\{1, 1, 50\}$ into $\{-1, -1, 50\}$ and $\{-1, 1, -50\}$.

- We can reconfigure the set $\{1, 2, 25\}$ into $\{-1, -2, 25\}$, $\{-1, 2, -25\}$, and $\{1, -2, -25\}$.

- We can reconfigure the set $\{1, 5, 10\}$ into $\{-1, -5, 10\}$, $\{-1, 5, -10\}$, and $\{1, -5, -10\}$.

- We can reconfigure the set $\{2, 5, 5\}$ into $\{-2, -5, 5\}$ and $\{2, -5, -5\}$.
- This makes 10 sets of two negative integers and one positive integer that multiply to 50, and four sets of three positive integers that multiply to 50. The answer is therefore 14 sets.

Be careful about where you apply generality. Applying it incorrectly will most likely result in undercounting.

Problems: Generality

1 Bronze. A long time ago, a factory was built directly off of a straight road that runs east to west. A car is somewhere far away from the factory, about to make a turn onto the road, and it can either make a left turn and begin going west or it can make a right turn and begin going east. What is the probability that it makes a turn in the correct direction?

2 Silver. A teacher is making three groups of three students out of a class of 9 students. What is the probability that Don and Nov are in Indov's group?

3 Gold. How many ways are there to arrange 3 identical yellow flowers and 2 identical red flowers in a circle? Rotations of the same arrangement are not considered distinct.

4 Gold. How many sets of three positive integers add up to 7?

5 Gold. How many ordered pairs of positive integers (a, b, c) add up to 6?

6 Gold. a, b, c and d are positive integers. If $a \times c = 3$, $b \times d = 6$, and $a^2 b + c^2 d = 29$, find one possible set of values of a, b, c and d.

7 Gold. Both x and y are positive integers. Find all ordered pairs (x, y) that satisfy $\dfrac{1}{x} + \dfrac{1}{y} = \dfrac{1}{6}$.

Solutions Manual

Part 1: Basic Probability

1 Bronze.

- A regular cubic dice has six faces numbered 1 through 6.
- There are six numbers that this dice can land on, and only one of them is 6.
- Therefore, the probability is $\frac{1}{6}$.

2 Bronze.

- There are two possible outcomes of a coin toss: landing on heads and landing on tails.
- One of these results in the coin landing on heads. Therefore, the probability is $\frac{1}{2}$.

3 Bronze.

- Let us first find the probability that he <u>will</u> choose either 32 or 74. Choosing 32 or 74 makes up two of the 100 possible choices, so the probability is $\frac{2}{100}$ or $\frac{1}{50}$.
- The probability that Vio will not pick 32 or 74 plus the probability that he will equals 1, so the final answer is $1 - \frac{1}{50}$ or $\frac{49}{50}$.

4 Bronze.

- 33 and 75 make up two of the 100 possible choices, so the probability is $\frac{2}{100}$ or $\frac{1}{50}$.

5 Silver.

- There are no positive integers between 14 and 16 that are less than 10. Since there are 0 successful outcomes, the probability is 0.

6 Silver.

- All of the positive integers between 14 and 17 are less than 20. Since the number of successful outcomes equals the number of total outcomes, the probability is 1.

7 Bronze.

- A total of $7 + 5 + 8 = 20$ marbles are in the bag. 5 of these marbles are blue.
- There are 5 successful choices and 20 possible choices, so the probability is $\frac{1}{4}$.

8 Bronze.

- The probability that she picks a blue marble is $\frac{5}{20}$, as shown in the previous problem.
- Since 1 minus the probability that an event will happen equals the probability that it will not happen (this is a slight variation of the original formula), the probability that the woman will not pick a blue marble is $1 - \frac{5}{20}$ or $\frac{3}{4}$.

9 Bronze.

- Since 75% or $\frac{3}{4}$ of the students in the classroom like blue, if a student is randomly selected out of the classroom, there is a $\frac{3}{4}$ chance that he or she likes blue.

10 Bronze.

- Similarly to the previous problem, if $\frac{1}{5}$ of the marbles in a bag are red, the probability that one randomly selected marble will be red is also $\frac{1}{5}$.

11 Silver.

- Since all of the numbers are distinct, there are seven possibilities as to which is the highest. All of these possibilities are equally likely.
- The girl's numbers constitute three of these seven possibilities, so the probability that her number is the highest is $\frac{3}{7}$.

12 Bronze.

- The number of possible choices is 6, as there are 6 elements in the set. Only one of the elements is 5, so there is 1 successful choice. Therefore, the probability is $\frac{1}{6}$.

13 Bronze.

- The area of the region that the dart must hit is 12 square units, and the area of the successful region is 2 square units.
- Therefore, the probability that a dart will hit the bulls-eye region is 2/12 or 1/6.

14 Silver.

- The carpet is 6 feet long. The ball has to stop somewhere along the 6 feet.
- The length of the successful outcome is 5–3 or 2 feet. Therefore, the probability is 2/6 or 1/3.

15 Silver.

- Only consider the day that Dino awakens.
- There are 24 hours in a day. The window of time that represents the successful outcome is 3 hours long. Therefore, the probability that the time will be between 8:00 a.m. and 11:00 a.m. is $\frac{3}{24}$ or $\frac{1}{8}$.
- The probability that an event will not occur equals 1 minus the probability that it will, so the probability that the time will not be between 8:00 a.m. and 11:00 a.m. is $1 - \frac{1}{8}$ or $\frac{7}{8}$.

Part 2: Basic Counting

1 Silver.

- If Kann turns left, he is faced with three paths. <u>Each</u> of the paths has five paths that lead to his destination, so there are 3×5 or 15 total paths that work when Kann turns left.

- If Kann turns right, there are only 3 pathways that he can take. Kann has to either turn right <u>or</u> turn left (he cannot do both), so we add 15 to 3 to make 18 paths in total.

2 Bronze.

- In the first room, there are 4 dolls per box for each of 50 boxes. Therefore, there are $50 \times 4 = 200$ dolls in the first room.

- In the second room, there are two dolls per box for each of 34 boxes. Therefore, there are $34 \times 2 = 68$ dolls in the second room.

- Adding the two, we find that there are $200 + 68 = 268$ dolls in total.

3 Silver.

- The boy is going to pick a shirt, a pair of shorts, <u>and</u> a pair of shoes. The number of ways he can do this is the number of distinct outfits he can make.

- Therefore, our answer is $5 \times 10 \times 2$ or 100 outfits.

4 Gold.

- The numbers that are going to have tens digits of 5 are 50 through 59, 150 through 159, 250 through 259, etc.

- There are 10 numbers from 50 to 59, but we also have to include 150–159, 250–259, ... up to 950–959. This makes a total of $10 + 10 \times 9 = 100$ numbers.

5 Silver.

- Let us first count the total number of outcomes. There are 11 books on the first desk and six on the second. If the woman is going to pick one book from the first desk and one book from the second, there are $11 \times 6 = 66$ ways that she can do this.

- Now let us count the number of successful outcomes. There are two English books on the first desk and three on the second.

There are $2 \times 3 = 6$ ways for the woman to pick an English book from the first desk and an English book from the second desk.

- The probability from what we now have is $\frac{6}{66}$ or $\frac{1}{11}$.

6 Bronze.

- There are 10 grams of sugar *for each* of the 20 liters. Therefore, there are 20×10 or 200 grams of sugar in total.

Part 3: Multiple Events

1 Bronze.

- The probability that the quarter will land heads up is $\frac{1}{2}$. The probability that the die will land on an even number is $\frac{3}{6}$ or $\frac{1}{2}$, since there are six possible numbers it could land on and three possible even ones.

- The probability that both of these will happen is $\frac{1}{2} \times \frac{1}{2}$, or $\frac{1}{4}$.

2 Silver.

- Since there are 10 special cards and 40 total cards, the probability that Ivonde will get a special card from the first stack is $\frac{10}{40}$ or $\frac{1}{4}$. The probability is the same for the second and third stacks.

- In the fourth stack, however, there are 30 cards that are not special and 40 total cards, so the probability of her getting a card that is not special is $\frac{30}{40}$ or $\frac{3}{4}$.

- $\frac{1}{4} \times \frac{1}{4} \times \frac{1}{4} \times \frac{3}{4} = \frac{3}{256}$.

3 Bronze.

- The probability that a single penny lands on heads is $\frac{1}{2}$. The probability that four pennies all land on heads is $\frac{1}{2} \times \frac{1}{2} \times \frac{1}{2} \times \frac{1}{2}$, or $\frac{1}{16}$.

4 Bronze.

- The probability that the first lamp that he picks will be blue is $\frac{4}{14}$, since there are 14 total lamps and 4 of them are blue.
- The probability that the second lamp he picks will be purple is $\frac{8}{14}$, since there are again 14 total lamps and 8 of them are purple.
- The probability that both will happen is $\frac{4}{14} \times \frac{8}{14}$ or $\frac{8}{49}$.

5 Silver.

- Each press of the button is an independent event. Their outcomes are unaffected by each other.
- Therefore, the probability that he presses the blue button on the third press is simply $\frac{1}{2}$.

6 Silver.

- The flavor that Ino buys does not matter. Whichever flavor he chooses, there is a $\frac{1}{5}$ probability that Vod will choose the same flavor, since there are 5 total flavors and only one of those is the right one.

7 Silver.

- The probability of a single person picking one of the two correct cups in a row of three is $\frac{2}{3}$.
- The probability of this happening in all four rows is $\frac{2}{3} \times \frac{2}{3} \times \frac{2}{3} \times \frac{2}{3}$ or $\frac{16}{81}$.

- $\dfrac{16}{81} \times \dfrac{16}{81}$ or $\dfrac{256}{6561}$ is the probability that two people will do this.

8 Bronze.

- The probability that a single dice lands on an even number is $\dfrac{3}{6}$ or $\dfrac{1}{2}$, since there are six numbers it could land on and three even ones.
- The probability that all 4 will land on an even number is $\dfrac{1}{2} \times \dfrac{1}{2} \times \dfrac{1}{2} \times \dfrac{1}{2}$ or $\dfrac{1}{16}$.

9 Gold.

- The probability that the event with probability q will not occur is $1 - q$.
- The probability of the first event occurring and second not occurring is $p \times (1 - q)$ or $p - pq$.

Part 4: Orderings

1 Bronze.

- Every different order of the digits 1, 2, 3, and 4 forms a different four-digit positive integer.
- The number of ways to order 4 distinct things is 4! or 24, so the number of four-digit positive integers that can be made with the digits 1, 2, 3, and 4 is also 24.

2 Bronze.

- Ino can draw the five jellybeans out of the jar in a total of 5! or 120 orders.
- Two specific orders allow her to win. Therefore, the probability that Ino wins is $\dfrac{2}{120}$ or $\dfrac{1}{60}$.

3 Silver.

- $29! = 29 \times 28 \times 27 \times \cdots \times 2 \times 1$.
- $28! = 28 \times 27 \times 26 \times \cdots \times 2 \times 1$.
- Imagine that the expression $\dfrac{29!}{28!}$ is fully expanded. $28 \times 27 \times 26 \times 25 \ldots \times 3 \times 2 \times 1$ will cancel in the numerator and denominator, leaving just 29.

4 Silver.

- If all of the digits were different, we could make 7! or 5040 positive integers by ordering them in different ways.
- However, there are three 2's, so we have to divide 7! by 3! or 6. Doing this, we find that the answer is 840 positive integers.

5 Bronze.

- Let us call the four chairs Chair 1, Chair 2, Chair 3, and Chair 4, respectively.
- There are four options for the cushion put on Chair 1, three for the cushion put on Chair 2 once one cushion is used up, two remaining for the cushion put on Chair 3, and one for the cushion put on Chair 4.
- This is $4 \times 3 \times 2 \times 1$ or 24 total ways to put all four cushions on a chair.

6 Silver.

- We first choose one spot in each row that is to be occupied by a coin. There are 3 ways to choose a spot in the first row, 4 ways to choose a spot in the second row, and 5 ways to choose a spot in the third row. There are $3 \times 4 \times 5 = 60$ ways to do this in total.
- Now we assign colors to the coins. The first coin can be red, yellow, or blue. This is 3 possibilities. 2 possibilities remain for the second coin, and only 1 possibility remains for the third coin after the second is assigned a color. Therefore, there are $3 \times 2 \times 1 = 6$ ways to assign colors to the coins.
- There are $60 \times 6 = 360$ ways to do both of these things.

7 Silver.

- If there were six cushions, there would be 6! or 720 ways to put them on the six chairs. However, three of the cushions are identical.

- Therefore, we divide 720 by 3! to obtain the answer, 120.

8 Platinum.

- In this problem, we use place value instead of finding all possible orders and adding them.

- There are 4! or 24 ways to order the digits 1, 2, 3, and 4 into different four-digit positive integers.

- How many of the positive integers have thousands digit 1? If 1 is assigned as the thousands digit, there are 3! or 6 ways to order the other three digits. Therefore, 6 of the total 24 positive integers have thousands digit 1.

- Similarly, 6 of the 24 positive integers have thousands digit 2, 6 have thousands digit 3, and 6 have thousands digit 4.

- In the sum of all 24 positive integers, each thousands digit is equivalent to 1000 times its value. Therefore, the six thousands digits of 1 add up to 6000, the six thousands digits of 2 add up to 12,000, and so forth. $6,000 + 12,000 + 18,000 + 24,000 = 60,000$.

- We now move on to the hundreds digits. How many of the 24 positive integers have hundreds digit 1? If 1 is assigned as the hundreds digit, there are 3! or 6 ways to order the other three digits. Going through the same process as before, we find that the hundreds digits sum up to $600 + 1200 + 1800 + 2400$ or 6000.

- Similarly, the tens digits sum up to 600 and the ones digits sum up to 60. Therefore, the final sum is 66,660.

Part 5: Dependent Events

1 Gold.

- We split this problem into three cases. The first case is if he wins on the first turn, the second case is if he wins on the second turn, and the third case is if he wins on the third turn. These cases must be exclusive.

- The probability that Vode wins on the first turn is $\frac{1}{6}$.

- The probability that Vode wins on the second turn is not so simple. For Vode to win on the second turn, he has to also not win on the first. The probability that Vode does not win on the first turn is $\frac{5}{6}$, and then the probability that he wins after this is $\frac{1}{6}$. The overall probability for this case is therefore $\frac{5}{6} \times \frac{1}{6}$, or $\frac{5}{36}$.

- Now we move on to case 3. The probability that Vode will not win on the first two turns is $\frac{5}{6} \times \frac{5}{6}$ or $\frac{25}{36}$, and the probability that he wins after this is $\frac{1}{6}$, so the total probability is $\frac{25}{36} \times \frac{1}{6}$ or $\frac{25}{216}$.

- Adding $\frac{1}{6}$, $\frac{5}{36}$, and $\frac{25}{216}$, we obtain the answer, $\frac{91}{216}$.

2 Silver.

- There are 15 total marbles in the jar, and 7 of them are red. Therefore, the probability that the first marble drawn will be red is $\frac{7}{15}$.

- After this marble is drawn, there are 6 red marbles and 14 total marbles left. The probability that the second marble drawn is red is $\frac{6}{14}$ or $\frac{3}{7}$. $\frac{7}{15} \times \frac{3}{7} = \frac{1}{5}$, our final answer.

3 Silver.

- There are 7 lamps in this collection, and 4 of them are purple. If two of the lamps are drawn without replacement, the probability that they will both be purple is $\frac{4}{7} \times \frac{3}{6}$ or $\frac{2}{7}$.

- There are also 7 total Frisbees in the collection, and 5 of them are yellow. If two Frisbees are drawn without replacement, the probability that they will both be yellow is $\frac{5}{7} \times \frac{4}{6}$ or $\frac{10}{21}$.

- Multiplying $\frac{10}{21}$ and $\frac{2}{7}$, we obtain our answer, $\frac{20}{147}$.

4 Bronze.

- The probability that Vode rolls a 3 and flips tails is $\frac{1}{6} \times \frac{1}{2}$ or $\frac{1}{12}$. The probability that Vode rolls a 6 and flips heads is again $\frac{1}{6} \times \frac{1}{2}$ or $\frac{1}{12}$.

- The probability that either one of these will happen is $\frac{1}{12} + \frac{1}{12}$ or $\frac{1}{6}$, since they are exclusive cases.

5 Gold.

- For Vode's arms to become sore on the second day, he has to not be sore on the first day and then become sore on the second day, since he'll only go to the gym on the second day if he is not sore.

- The probability that he will not become sore on the first day is $1 - \frac{2}{3}$ or $\frac{1}{3}$, and the probability that he will get sore given that he makes it to the second day is $\frac{2}{9}$. Therefore, the total probability is $\frac{1}{3} \times \frac{2}{9}$ or $\frac{2}{27}$.

6 Gold.

- A given child liking blue and the same child liking red are completely independent events. Therefore, the probability that a child will like red given that he/she likes blue equals the probability that any child will like red, since they are not affected by each other.

- What is the probability that a given child will like red? The probability that the child will like blue is 80% or $\frac{4}{5}$, and the probability that the child will like both blue and red is 25% or $\frac{1}{4}$. Let us call the probability we are looking for x.

- We set up the equation $\frac{4}{5} \times x = \frac{1}{4}$. Solving, we find that $x = \frac{5}{16}$.

7 Gold.

- Let us differentiate the first and second choices (this means that picking one square and then another is different from picking the second square and then the first). We are able to differentiate the choices when calculating probability if we differentiate both the number of successful and the number of total outcomes. The probability of picking two squares that share a vertex disregarding order equals the probability of picking two squares that share a vertex considering the order in which they were chosen because the order of the squares has nothing to do with whether the two squares share a vertex.

- There are 16 ways to pick the first square and 15 ways to pick the second, making a total of 16×15 or 240 ways to pick the two squares.

- Now we count the number of successful outcomes. If one of the four corner squares are picked first, there are three squares that share a vertex with it. Therefore, there are 4×3 or 12 successful outcomes in this case.

- If one of the four middle squares is picked first, there are eight squares that share a vertex with it. There are 4×8 or 32 successful outcomes in this case.

- If one of the 8 squares on the edges of the larger square but not in the corners is picked first, there are five squares that share a vertex with it, so there are 8×5 or 40 successful outcomes.

- The total number of successful outcomes is $12 + 32 + 40$ or 84, and the final probability is $\dfrac{84}{240}$ or $\dfrac{7}{20}$.

8 Silver.

- Let us label the possible drinks Drink 1, Drink 2, and Drink 3, respectively. For all four friends to receive the same drink, all receive Drink 1, all receive Drink 2, or all receive Drink 3.

- The probability that all of the friends receive Drink 1 is $\dfrac{1}{3} \times \dfrac{1}{3} \times \dfrac{1}{3} \times \dfrac{1}{3} = \dfrac{1}{81}$, and this probability is the same for Drinks 2 and 3 as well.

- Adding the probabilities of the cases, we find that the probability that all four friends receive the same drink is $\dfrac{1}{27}$.

9 Gold.

- The only alternating sequence of purple marbles and blue marbles with five blue marbles and four purple marbles is B-P-B-P-B-P-B-P-B. Therefore, the answer to this problem is the probability that Ino will obtain this sequence of marbles.

- The probability that she will pick a blue first is $\dfrac{5}{9}$, the probability that she will pick a purple next is $\dfrac{4}{8}$, the probability that she will pick a blue third is $\dfrac{4}{7}$, the probability that she will pick a purple fourth is $\dfrac{3}{6}$, and then $\dfrac{3}{5}, \dfrac{2}{4}, \dfrac{2}{3}, \dfrac{1}{2}$, and 1 respectively for the last five marbles.

- To calculate the final probability, we multiply all of the individual probabilities, obtaining $\dfrac{1}{126}$ as our answer.

10 Platinum.

- This is not a truly fair game. Since she goes first, the girl has a slight advantage over the boy.

- We use casework to solve this problem. The probability that the girl wins in her first turn is $\frac{1}{9}$, since there are 9 total cards and one of them is the winning one.

- Now we move on to the probability that the girl wins on her second turn. For this to happen, the girl must not win on her first turn, the boy must not win on his first turn, and then the girl must win after this. The probability that the girl wins on her second turn is $\frac{8}{9} \times \frac{8}{9} \times \frac{1}{9}$.

- We use this same process to find the probability that the girl wins on her third turn: this probability turns out to be $\frac{8}{9} \times \frac{8}{9} \times \frac{8}{9} \times \frac{8}{9} \times \frac{1}{9}$ The probability of her winning on her fourth turn is $\frac{8}{9} \times \frac{8}{9} \times \frac{8}{9} \times \frac{8}{9} \times \frac{8}{9} \times \frac{8}{9} \times \frac{1}{9}$, the probability of her winning on her fifth turn is $\frac{8}{9} \times \frac{8}{9} \times \frac{8}{9} \times \frac{8}{9} \times \frac{8}{9} \times \frac{8}{9} \times \frac{8}{9} \times \frac{8}{9} \times \frac{1}{9}$, and so forth. This game could technically go on forever.

- If we add up all of the cases to complete the casework, we will end up with $\frac{1}{9} + \left(\frac{8}{9} \times \frac{8}{9} \times \frac{1}{9}\right) + \left(\frac{8}{9} \times \frac{8}{9} \times \frac{8}{9} \times \frac{8}{9} \times \frac{1}{9}\right) + \left(\frac{8}{9} \times \frac{8}{9} \times \frac{8}{9} \times \frac{8}{9} \times \frac{8}{9} \times \frac{8}{9} \times \frac{1}{9}\right) \ldots$ Do you recognize what this is? This is an infinite series! The first term is $\frac{1}{9}$, and the common ratio is $\frac{8}{9} \times \frac{8}{9}$ or $\frac{64}{81}$. Therefore, the sum of this series is $\frac{1}{9} / \left(1 - \frac{64}{81}\right)$, which equals $\frac{9}{17}$.

11 Gold.

- The odd positive integers between 1 and 100 inclusive are 1, 3, 5, 7, 9, ..., 97, 99. This makes 50 numbers. The even positive integers between 1 and 100 inclusive are 2, 4, 6, 8, 10, ..., 96, 98, 100. This is also a total of 50 numbers.

- To solve this problem, we must realize that the product of two odd numbers is odd, the product of an odd number and an even number is even, and the product of two even numbers is even.

- Therefore, if the person picks an even number as his or her first pick, the second number could be anything and the product will still be even. If the person picks an odd number as his or her first pick, the second pick must be even for the product to be even.

- Now we do a little bit of casework. The probability that the first pick is even is $\frac{50}{100}$ or $\frac{1}{2}$. The probability that the pick after that makes an even product is 1 (guaranteed), so the total probability for this case is $\frac{1}{2} \times 1$ or $\frac{1}{2}$.

- The probability that the person's first choice is odd is $\frac{1}{2}$. The probability that the person's second choice is even is $\frac{1}{2}$ as well. Therefore, the total probability for this case is $\frac{1}{2} \times \frac{1}{2}$ or $\frac{1}{4}$. Finally, we add the probabilities of the cases, finding that the answer is $\frac{3}{4}$.

12 Silver.

- If Vio sits on one of the end seats, there is only one seat next to him, but if Vio sits in one of the three middle seats, there are two seats next to him.

- There is a $\frac{2}{5}$ chance that Vio takes one of the end seats, and there is a $\frac{3}{5}$ chance that he takes one of the middle ones.

- If he takes one of the end seats, there is a $\frac{1}{4}$ chance that Oniv will take the one seat next to him, as there are only four seats remaining for her to take. If he takes one of the middle seats, there is a $\frac{2}{4}$ or $\frac{1}{2}$ chance that Oniv will take one of the two seats next to him.

- Therefore, the total probability that Vio and Oniv will sit next to each other is $\left(\frac{2}{5} \times \frac{1}{4}\right) + \left(\frac{3}{5} \times \frac{1}{2}\right)$, which equals $\frac{2}{5}$.

- Notice that since there was no reason to take into account what the other three people were doing, we ignored the fact that they were in the problem.

13 Silver.

- There are two ways for this condition to be satisfied. One is if we receive a heads and then a tails. The other is if we receive a tails and then a heads. The probability of receiving a heads and then a tails is $\frac{2}{3} \times \frac{1}{3} = \frac{2}{9}$. The probability of receiving a tails and then a heads turns out to be the same.

- The last step is adding these probabilities, since both cases satisfy the requirement. Doing this yields $\frac{4}{9}$, our answer.

14 Gold.

- If Ino's two numbers are both 25 or lower, there is a 100% chance that Node's number will be the highest.

- If one of Ino's numbers is between 26 and 50 inclusive, it is not guaranteed that Node's number will be the highest, and if both of her numbers are between 26 and 50 inclusive, there is an even smaller chance that Node's number will.

- Let us examine the first case, in which both of Ino's numbers are 25 or lower. On Ino's first pick, there is a $\frac{25}{50}$ or $\frac{1}{2}$ probability that she picks a number that is 25 or lower, as the successful outcomes are the picks 1–25 and the total number of outcomes is 50. On Ino's second pick, there is a $\frac{24}{49}$ chance of her picking another number that is 25 or lower, as she has already picked one. The probability that this case happens is therefore $\frac{1}{2} \times \frac{24}{49}$ or $\frac{12}{49}$, but once this happens the probability that Node's number will be the highest is 1.

- Next, we examine the second case, in which one of Ino's numbers is between 26 and 50 and one of her numbers is not. The probability that this case occurs is $\frac{1}{2}$, as the probability that she picks a low number first and then a high number is $\frac{1}{2} \times \frac{1}{2}$ or $\frac{1}{4}$, and the probability that she picks a high number first and then a low number is the same. Once this happens, there is a $\frac{1}{2}$ chance that Node's number will be the highest, as they both have different numbers in the 26–50 range, so there must be a 50–50 chance that each of their numbers will be the highest.

- We now examine the last case, in which both of Ino's numbers are in the 26–50 range. We have already found the probability that her numbers are both in the 1–25 range, and it should be exactly the same for this case. Therefore, the probability that this will happen is also $\frac{12}{49}$. Once this happens, there is a $\frac{1}{3}$ chance that Node's number will be the highest, as there is an equal chance of being the highest for all three numbers, and only one of them is Node's.

- Our final step is completing the casework. $\frac{12}{49}(1) + \frac{1}{2}\left(\frac{1}{2}\right) + \frac{12}{49}\left(\frac{1}{3}\right) = \frac{113}{196}$.

15 Gold.

- To solve this problem, we must split it up into two cases. The first case is if event A occurs but event B does not, and the second case is if event A does not occur but event B does.

- The probability that event A occurs is p, and the probability that event B does not is $1 - q$. Therefore, the total probability for this case is $p(1 - q)$, which equals $p - pq$.

- Using a similar method, we find that the probability of the second case is $q(1 - p)$ or $q - pq$. Adding the probabilities of the two exclusive cases yields $p + q - 2pq$, our answer.

16 Silver.

- Before finding the probability that the basketball team can practice, we find the probability that they cannot practice.

- In the previous problem, we found that given independent events A and B with respective probabilities of p and q, the probability of either A or B occurring but not both is $p + q - 2pq$.

- The probability of both p and q occurring is pq, so, by casework, the probability of either p or q or both occurring is $p + q - 2\,pq + pq$ or $p + q - pq$. Therefore, the probability that the janitor forgets to unlock the gym or the volleyball team wants to practice or both is $\dfrac{1}{500} + \dfrac{1}{3} - \dfrac{1}{1500}$, which ends up being equivalent to $\dfrac{502}{1500}$ or $\dfrac{251}{750}$. Our final answer is therefore $1 - \left(\dfrac{251}{750}\right) = \dfrac{499}{750}$.

17 Silver.

- The probability that at least one of the candies is yellow is equivalent to 1 minus the probability that none of them are yellow, since the probability that something happens plus the probability that it does not happen equals 1.

- The probability that the first pick is not yellow is $\dfrac{10}{14}$, and the probability that the second pick is not yellow is $\dfrac{9}{13}$. Therefore, $\dfrac{10}{14} \times \dfrac{9}{13}$ or $\dfrac{45}{91}$ is the probability that neither pick is yellow. $1 - \dfrac{45}{91} = \dfrac{46}{91}$.

Part 6: Subsets

1 Bronze.

- Since this set has 4 distinct elements, the number of subsets of the set is 2^4 or 16.

2 Bronze.

- This set has 5 elements, so the number of subsets of the set is 2^5 or 32.

- However, this includes the one subset that has no elements, the empty subset. Therefore, there are 31 non-empty subsets.

3 Silver.

- This set has two 1's, two 2's, three 3's, one 4, and one 5.

- In any subset, you can have no 1's, one 1, or two 1's. This is three options for the 1's.

- Similarly, there are three options for the 2's, four for the 3's, two for the 4's, and two for the 5's. Therefore, the total number of subsets is $3 \times 3 \times 4 \times 2 \times 2 = 144$ subsets.

4 Silver.

- Onid can purchase no fans, one fan, or two fans.

- He can purchase no tables, one table, or two tables.

- Lastly, he can purchase either no chairs or one chair. Therefore, there are $3 \times 3 \times 2 = 18$ combinations of items that he can purchase.

5 Gold.

- To solve this problem, we first find the number of ways spices can be added to the soup and the number of ways sweeteners can be added to the soup, and then multiply the two to find the final answer.

- The number of ways that we can add spices to the soup equals the number of non-empty subsets of a four element set. This is because we can add any subset of the four different spices that we choose to the soup, except the empty subset.

- Therefore, there are $2^4 - 1 = 15$ ways to add spices to the soup. Similarly, there are $2^3 - 1 = 7$ ways to add sweeteners to the soup.

- We are adding both spices and sweeteners to the soup, so there are $15 \times 7 = 105$ ways to do this.

Part 7: Organized Counting

1 Silver.

- There are two places where a 9 can be in a number from 1 to 99: the units digit and the tens digit.

- In the 10 numbers 9, 19, 29, 39, 49, 59, 69, 79, 89, and 99, I will have to write a 9 in the units digit.

- In the 10 numbers 90 through 99, I will have to write a 9 in the tens digit. Therefore, I will write the digit 9 a total of 20 times.

2 Bronze.

- We start with the one-element subsets: $\{w\}$, $\{x\}$, $\{y\}$, $\{z\}$.

- Now for the two-element subsets: $\{w, x\}$, $\{w, y\}$, $\{w, z\}$, $\{x, y\}$, $\{x, z\}$, $\{y, z\}$.

- Now for the three-element subsets: $\{w, x, y\}$, $\{w, x, z\}$, $\{w, y, z\}$, $\{x, y, z\}$.

- Finally, the four-element subset: $\{w, x, y, z\}$.

3 Gold.

- Let us count in order, starting with where the hundreds digit is 1.

- There are 8 numbers here whose digits add up to 12: 129, 138, 147, 156, 165, 174, 183, and 192.

- Similarly counting the numbers with hundreds digits of 2, 3, 4, 5, 6, 7, 8, and 9, we find that there are 9 numbers with hundreds digit 2 whose digits add up to 12, 10 with hundreds digit 3, 9 with hundreds digit 4, 8 with hundreds digit 5, 7 with hundreds digit 6, 6 with hundreds digit 7, 5 with hundreds digit 8, and 4 with hundreds digit 9.

- That is $8 + 9 + 10 + 9 + 8 + 7 + 6 + 5 + 4$ or 66 integers.

4 Bronze.

- Each dice has six possible outcomes: 1, 2, 3, 4, 5, and 6. We must count all of the combinations that add up to 7.

- We start with 1 on the first dice and 6 on the second, then 2 and 5, 3 and 4, 4 and 3, 5 and 2, and lastly 6 and 1. This is a total of 6 successful outcomes.

- The total number of outcomes is 6×6 or 36. Therefore, the probability is $6/36$ or $1/6$.

5 Silver.

- We start by counting the cases where all the bags have the same number of marbles. He can pick 3 3-marble bags, 3 5-marble bags, or 3 7-marble bags, leaving him with 9 marbles, 15 marbles, or 21 marbles, respectively.

- We now continue to the cases with 2 types of bags: if he picks out 3-marble bags and 5-marble bags, he can receive $3 + 3 + 5$ (11) or $3 + 5 + 5$ (13) marbles, if he picks out 3-marble bags and 7-marble bags, he can receive $3 + 3 + 7$ (13) or $3 + 7 + 7$ (17) marbles, and if he picks out 5-marble and 7-marble bags, he can receive $5 + 5 + 7$ (17) or $5 + 7 + 7$ (19) marbles.

- Lastly, if he picks out 3 distinct types of bags, there is only one sum that he can receive: $3 + 5 + 7$ or 15 marbles. This makes 7 *distinct* numbers of marbles he can obtain: 9, 11, 13, 15, 17, 19, and 21.

6 Silver.

- If we start by assigning the first two digits of a four-digit number, then set the last digit as the first and the third digit as the second, we create a palindrome. Therefore, by counting the number of ordered sets of two even integers that form the first two digits of the four-digit integer, we count the number of even four-digit palindromes, since each of these two-digit sets produces a distinct four-digit palindrome.

- There are four possibilities for the first digit—2, 4, 6 and 8— and there are five possibilities for the second—0, 2, 4, 6, and 8.

- Therefore, there are 20 possibilities for the first two digits, and consequently 20 four-digit palindromes with no odd digits.

7 Silver.

- What if Vode makes 0 3-point shots? If he makes 18 2-point shots, he ends with 36 points.

- If he makes 1 or any odd number of 3-point shots, there will be no way for him to make it an even 36 points.

- However, if he makes any even number from 0 to 12 of them, he will be able to make a certain number of two-point shots to fill the gap. There are 7 even numbers from 0 to 12 including both 0 and 12, so there are 7 ways for Vode to score 36 points.

8 Silver.

- Since we are selecting two numbers, let us go in order of the first number. If we select 1 first, we need 100 to make the product 100, so this does not work.

- If we select 2 first, selecting 50 yields a product 100.

- If we select 4 first, selecting 25 yields a product of 100.

- If we select 5 first, selecting 20 yields a product of 100.

- If we select 10 first, selecting another 10 yields a product of 100.

- The next number we can pick is 20, but we already counted the pair with 20 when we selected 5 first. Therefore, we are done with the 4 pairs 2 and 50, 4 and 25, 5 and 20, and 10 and 10.

9 Silver.

- Let us count in order of the first positive integer.

- The sets of positive integers that add up to 50 are 1 and 49, 2 and 48, 3 and 47, 4 and 46, and so forth until 49 and 1. We know that this comprises to 49 sets because the first integer starts at 1 and increases by 1 until it reaches 49, and each increase by 1 corresponds to one pair.

10 Gold.

- There is more than 1 way to split this problem up into different cases. However, we will split this problem into the cases where there are no quarters, 1 quarter, 2 quarters, and 3 quarters.

- Case 1: If there are no quarters, $0.85 can be made with 8 dimes and 1 nickel, 7 dimes and 3 nickels, 6 dimes and 5 nickels, and so forth up to 0 dimes and 17 nickels. This makes 9 possibilities.

- Case 2: If there is 1 quarter, we need to make $0.60 in dimes and nickels. This can be done with 6 dimes and no nickels, 5 dimes and 2 nickel, 4 dimes and 4 nickels, all the way up to 0 dimes and 12 nickels. This makes 7 possibilities.

- Case 3: If there are two quarters, we need to make $0.35 in dimes and nickels. This can be done with 3 dimes and 1 nickel, 2 dimes and 3 nickels, 1 dime and 5 nickels, or 0 dimes and 7 nickels. This makes 4 possibilities.

- Case 4: If there are 3 quarters, we need to make $0.10 in dimes and nickels. We can do this with either 0 dimes and 2 nickels or 1 dime and no nickels. This makes two possibilities.

- We counted a total of $9 + 7 + 4 + 2$ or 22 ways to make $0.85.

11 Platinum.

- We want the greatest positive integer in this set to be as large as possible, and since there is a limit on the sum of all seven elements in the set, we therefore have to make the sum of the first six as small as possible.

- Since 5 is the unique mode, there has to be more than one 5. Let us first see how small we can get the sum of the first six positive integers with two 5's. Since 5 is the unique mode, we cannot repeat any numbers. This leaves 1, 2, 3, 4, 5, 5 as the sequence of six positive integers with the least possible sum. $1 + 2 + 3 + 4 + 5 + 5 = 20$.

- Now let us see how low we can go with three 5's. With three 5's, we can repeat numbers up to twice and still keep 5 as the unique mode. 1, 1, 2, 5, 5, 5, is the sequence with the lowest possible sum in this case. $1 + 1 + 2 + 5 + 5 + 5 = 19$.

- If we move on to four 5's, without adding anything else, the sum of the four 5's is 20. 20 is greater than our lowest possible sum so far, so we can stop now.

- Since the mean of the seven positive integers is 8, they add up to 56. We know that 1, 1, 2, 5, 5, 5 are the first six positive integers that still satisfy the requirements of the problem but yield the smallest sum. Therefore, the greatest positive integer that can be in this set is $56 - 19$ or 37.

12 Silver.

- The number of picks that guarantees 3 yellow marbles is the maximum number of picks that it can possibly take to obtain 3 yellow marbles.

- What is this worst case scenario? The worst case scenario is if the person picks out all of the candies except for the yellow ones, and then has to pick 3 yellow candies to satisfy his requirement.

- There are 18 candies that are not yellow in the jar, so the worst case scenario requires the person to pick $18 + 3$ or 21 candies out of the jar. Therefore, picking 21 candies out of the jar guarantees at least three yellow candies because there is no way that doing this will not yield at least three.

Part 8: Permutations and Combinations

1 Bronze.
- $^9P_3 = \dfrac{9!}{(9-3)!} = \dfrac{9!}{6!} = 9 \times 8 \times 7 = 504.$

2 Bronze.
- $^6C_4 = \dfrac{6!}{4! \times (6-4)!} = \dfrac{6!}{4! \times 2!} = \dfrac{6 \times 5}{2!} = 15$ (Since $6! = 6 \times 5 \times 4!$).

3 Silver.

- 95 choose 94 $= \dfrac{95!}{94!\,(95-94)!} \cdot \dfrac{95!}{94!}$ simplifies to just 95, since $94 \times 93 \times 92 \times \cdots \times 2 \times 1$ cancels on both top and bottom. This leaves us with $\dfrac{95}{1!}$, which equals 95. This logic can be used to prove the universal rule that $\binom{n}{m} = \binom{n}{n-m}$.

4 Bronze.

- The number of pairs of garden gnomes that the boy can purchase is the number of ways that he can choose two garden gnomes out of the 20 where the order they are chosen in does not matter.
- This equals 20 choose 2, or $\dfrac{20 \times 19}{2}$. Simplifying yields 190 pairs.

5 Silver.

- The number of ways that she can make a stack of four bricks is the number of ways that she can choose four bricks from the set of 12 where order does matter.
- This is because a stack is made up of four bricks in the order they were chosen.
- Therefore, the answer is 12 permute 4 or 11,880 stacks.

6 Silver.

- The probability that one or both of the cards she picks will have a blue dot equals one minus the probability that neither of the cards will have one.
- The number of ways that Vio can pick any two cards is 22 choose 2 or 231.
- The number of ways Vio can pick two cards without blue dots is 17 choose 2 or 136, since there are 17 cards without blue dots in the jar.
- Therefore, $\dfrac{136}{231}$ is the probability that both of Vio's cards will not have a blue dot.
- $1 - \dfrac{136}{231} = \dfrac{95}{231}$ is our answer.

7 Silver.

- The number of three-element subsets of a 13-element set is equivalent to the number of ways to choose three distinct elements out of the 13 elements.
- It is now clear that the answer is $\binom{13}{3} \cdot \binom{13}{3} = \dfrac{13 \times 12 \times 11}{3 \times 2}$ $= 286$ three-element subsets.

8 Silver.

- Although they are doing so in a very indirect way, the group of people is simply choosing 3 hats out of the 25.
- The number of ways that they can do this is 25 choose 3, or $\dfrac{25 \times 24 \times 23}{3!}$. This simplifies to 2300 sets of three hats.

9 Gold.

- Out of the 11 flips, three of them need to land tails. How many sequences of heads and tails satisfy this? One such sequence is HHHHHHHHTTT.
- To account for all possible sequences, we choose 3 out of the 11 flips to be tails. This can also be thought of as choosing 3 out of the 11 spots in a sequence of 8 H's and 3 T's to be T. $\binom{11}{3} = \dfrac{11 \times 10 \times 9}{3!} = 165.$
- Each one of these sequences, for example, HHHHHHHHTTT, has a probability of $\left(\dfrac{1}{2}\right)^{11}$ or $\dfrac{1}{2048}$. The sum of the probabilities of the cases equals $\dfrac{1}{2048} + \dfrac{1}{2048} + \dfrac{1}{2048} \cdots$ (165 times), which is $\dfrac{1}{2048} \times 165$ or $\dfrac{165}{2048}$.

10 Gold.

- To solve this problem, we first choose 3 out of the 6 squares to be colored. There are $\binom{6}{3}$ or 20 ways to do this.
- Next, we assign each of the 3 squares we chose a different color. The same color cannot be used twice, so there are 3 possible colors for the first square, 2 possible colors for the

second, and one for the third. Therefore, there are $3 \times 2 \times 1$ or 6 ways to color the 3 chosen squares.

- Each of the three remaining squares can be either colored black or left white. Therefore, there are $2 \times 2 \times 2$ or 8 ways to color the squares that are not being colored with the markers.

- This is a total of $20 \times 6 \times 8$ or 960 distinct ways to color the grid.

11 Silver.

- The number of ways to pick 3 positive integers out of the first 15 is 15 choose 3, which equals $\dfrac{15 \times 14 \times 13}{3!}$ or 455.

- The person cannot pick 8 or 13 to obtain a successful outcome, so there are 13 numbers left that she can pick. Therefore, the total number of successful outcomes is 13 choose 3, which equals $\dfrac{13 \times 12 \times 11}{3!}$ or 286.

- $\dfrac{286}{455}$ can be simplified into $\dfrac{22}{35}$ by dividing both top and bottom by 13.

12 Gold.

- Let us consider one sequence of heads and tails that satisfies this problem: TTTTHHHH. The probability of obtaining this exact sequence is $\dfrac{1}{3} \times \dfrac{1}{3} \times \dfrac{1}{3} \times \dfrac{1}{3} \times \dfrac{2}{3} \times \dfrac{2}{3} \times \dfrac{2}{3} \times \dfrac{2}{3}$, or $\dfrac{16}{6561}$. All other sequences that satisfy the problem, for example, THTHTTHH, also have the same probability due to the Commutative Property of Multiplication.

- How many of these sequences are possible? The number of ways to order 8 things where two sets of 4 are repeated is $\dfrac{8!}{4! \times 4!}$, which equals $\dfrac{8 \times 7 \times 6 \times 5}{4 \times 3 \times 2 \times 1}$ or 70. This can also be thought of as choosing 4 spots of the 8 to be heads and leaving the rest as tails. $\binom{8}{4}$ also equals 70.

- If we consider each sequence a case, there are 70 cases, each with probability $\dfrac{16}{6561}$. The sum of their probabilities is $\dfrac{16}{6561} \times 70$ or $\dfrac{1120}{6561}$.

13 Gold.

- If there were x people at the gathering, how many phone numbers were exchanged in terms of x? The answer is $\binom{x}{2}$, since the number of ways to choose two people to exchange phone numbers is equivalent to the number of phone numbers exchanged. This is because everyone exchanged phone numbers with everyone else exactly once.

- To find x, we set up the equation $\binom{x}{2} = 153$. $\binom{x}{2} = \dfrac{x(x-1)}{2}$, and $\dfrac{18 \times 17}{2} = 153$.

- Therefore, there were 18 people at the gathering.

14 Gold.

- There are 2^n subsets of a set with n elements.

- How many zero-element subsets does a set with n elements have? This is equivalent to the number of ways there are to choose 0 elements out of the n elements, which is $\binom{n}{0}$.

- Similarly, the number of one-element subsets of a set with n elements is $\binom{n}{1}$, the number of two-element subsets of a set with n elements is $\binom{n}{2}$, and so on.

- The number of zero-element subsets of a set plus the number of one-element subsets of the set plus the number of two-element subsets and so forth up to the number of n-element subsets of the set (n the maximum number of elements in a subset) equals the total number of subsets of the set.

- This proves the assertion that $\binom{n}{0} + \binom{n}{1} + \binom{n}{2} + \binom{n}{3} + \ldots + \binom{n}{n-1} + \binom{n}{n} = 2^n$.

15 Silver.

- By the Binomial Theorem, $(a + b)^{11} = \binom{11}{0} \times a^{11} + \binom{11}{1} \times a^{10}b + \binom{11}{2} \times a^9 b^2 \ldots$

- $\binom{11}{0} = 1, \binom{11}{1} = 11$, and $\binom{11}{2} = \dfrac{11 \times 10}{2!} = 55$

- Therefore, the first three terms of $(a + 3b)^{11}$ when in standard order are a^{11}, $11 \times a^{10} \times 3b$ or $33a^{10}b$, and $55 \times a^9 \times (3b)^2$ or $495a^9 b^2$.

Part 9: Expected Value

1 Gold.

- 1 can occur 0, 1, 2, or 3 times. The probability of rolling no 1's is $\dfrac{5}{6} \times \dfrac{5}{6} \times \dfrac{5}{6}$ or $\dfrac{125}{216}$, since in each dice, there are 6 possible outcomes in total and 5 possible outcomes that are not 1.

- The probability of rolling one 1 is $\dfrac{1}{6} \times \dfrac{5}{6} \times \dfrac{5}{6} \times 3$ or $\dfrac{75}{216}$; the $\times 3$ is due to the fact that any of the three dice can be the one that lands on 1.

- The probability of rolling two 1's is $\dfrac{1}{6} \times \dfrac{1}{6} \times \dfrac{5}{6} \times 3$ or $\dfrac{15}{216}$; the $\times 3$ here is due to the fact that any of the three dice can be the one that does not land on 1.

- Lastly, the probability of rolling three 1's is $\dfrac{1}{6} \times \dfrac{1}{6} \times \dfrac{1}{6}$ or $\dfrac{1}{216}$.

- Therefore, the expected number of occurrences of 1 is $\dfrac{125}{216} \times 0 + \dfrac{75}{216} \times 1 + \dfrac{15}{216} \times 2 + \dfrac{1}{216} \times 3$, which equals $\dfrac{1}{2}$.

2 Gold.

- There are 27 total coins in the bag. The probability of choosing a quarter is $\dfrac{12}{27}$ or $\dfrac{4}{9}$, the probability of choosing a dime is $\dfrac{10}{27}$, and the probability of choosing a penny is $\dfrac{5}{27}$.

- If Viod chooses a quarter, he obtains 25 cents, if he chooses a dime, he obtains 10 cents, and if he chooses a penny, he obtains 1 cent.
- Therefore, his expected earnings are $\frac{4}{9} \times 25 + \frac{10}{27} \times 10 + \frac{5}{27} \times 1$ cents, which equals $0.15.

3 Silver.

- If each player has an equal chance of winning each match, it follows that each player has an equal chance of winning the whole tournament.
- Since there are 8 players in the tournament, and one of them must win, the probability that a given player will win the tournament and receive the $24 is $\frac{1}{8}$. If the player does not win (this has a $\frac{7}{8}$ probability), he or she will receive $0.
- Therefore, the expected earnings are $\frac{1}{8} \times 24 + \frac{7}{8} \times 0 = \$3$.

4 Silver.

- By the definition of probability, Dinov is expected to win one in every three games he plays.
- We set up the proportional equation $\frac{1 \text{ win}}{3 \text{ games played}} = \frac{x \text{ wins}}{24 \text{ games played}}$. Solving, we find that $x = 8$ wins.

Part 10: Weighted Average

1 Silver.

- There are $200 + 100 + 100 = 400$ points in total. The 200-point test's weight as a fraction is $\frac{200}{400}$ or $\frac{1}{2}$, and the two 100-point tests' weights are each $\frac{100}{400}$ or $\frac{1}{4}$.

- The weighted average of the test scores is therefore $\left(\frac{1}{2} \times 93\right) + \left(\frac{1}{4} \times 89\right) + \left(\frac{1}{4} \times 78\right) = 46.5 + 22.25 + 19.5 = 88.25\%$.

2 Silver.

- Each price has a respective fraction of occurrence as to how many vases out of the total cost that amount.

- The fraction of occurrence of \$1000 is $\frac{1}{2}$, the fraction of occurrence of \$2000 is $\frac{1}{4}$, the fraction of occurrence of \$3000 is $\frac{1}{5}$, and the fraction of occurrence of \$500 is $\frac{1}{20}$. Therefore, the weighted average of the prices is $\frac{1}{2} \times 1000 + \frac{1}{4} \times 2000 + \frac{1}{5} \times 3000 + \frac{1}{20} \times 500 = \$1625$

3 Platinum.

- The average speed of the car is the weighted average of all of the speeds it travels at during its trip. Each one of these speeds can be represented as a distance divided by a time.

- Let us call one of the speeds that the car travels at s_1. Let us call the distance that it travels at this speed d_1 and the time that it travels for at this speed t_1. (Do not worry about the subscripts, they just help label the variables). We know that $s_1 = \frac{d_1}{t_1}$.

- When calculating the weighted average, we multiply the speed by its fraction of occurrence, which in this case is the time that it travels at the certain speed out of the total time. The speed s_1's fraction of occurrence is $\frac{t_1}{t_{\text{total}}}$, where t_{total} is the total amount of time that the car travels.

- When we multiply this fraction by $\frac{d_1}{t_1}$, the result is $\frac{d_1}{t_{\text{total}}}$. When we add all the different parts of the weighted average $\left(\frac{d_2}{t_{\text{total}}}, \frac{d_3}{t_{\text{total}}}, \text{ etc.}\right)$, we are left with the total distance over the total time. This is the car's average speed.

- If we multiply by the total time, we end up with the total distance that the car travels.

Part 11: Generality

1 Bronze.

- We know that the car is not exactly at the factory's position. No matter what the position of the car is, exactly one of the two possible ways that it can turn is the correct one. Both are equally likely. Therefore, the probability that it makes a turn in the correct direction is $\frac{1}{2}$.

2 Silver.

- Let us call Indov's group Group Indov. The total number of ways to choose two other students to be in Indov's group is $\binom{8}{2}$ or 28.

- Therefore, the probability that the two other students in Indov's group will be Don and Nov is $1/28$.

3 Gold.

- If we put the 2 identical red flowers into two spots, the 3 yellow flowers will fit in into the three remaining spaces. Therefore, we only have to worry about arranging the 2 red flowers.

- Let us label the spots on the circle 1, 2, 3, 4, and 5. Without loss of generality, let us assign the first red flower to spot 1. The second red flower can now go into any of the remaining four spots. This yields four possible arrangements.

- However, if the second red flower goes into spot 4, the resulting arrangement can be rotated such that it is identical to the arrangement where the second red flower goes into spot 2.

- The same thing can be done to the arrangement where the second red flower goes into spot 5; it can be rotated such that it is identical to the arrangement where the second red flower goes into spot 1.

- Therefore, there are only 2 possible arrangements.

4 Gold.

- Let us call the three positive integers a, b, and c. Without loss of generality, let us make $a \leq b \leq c$.
- a has to be either 1 or 2, because if a is 3, the minimum possible sum is 9.
- From here, it is easy to count the possible sets. They are 1 1 5, 1 2 4, 1 3 3, and 2 2 3. That makes 4 sets of positive integers.

5 Gold.

- Our strategy here will be to find the number of unordered sets of three positive integers that add up to 6, and then from there find the number of ordered triples that can be made out of each of the unordered sets.
- Without loss of generality, let us make $a \leq b \leq c$.
- a has to be either 1 or 2, otherwise the minimum sum will be too high. From here, it is easy to count the possible outcomes. They are 1 1 4, 1 2 3, and 2 2 2.
- How many ordered triples can be formed out of 1 1 4? This is equivalent to the number of ways to order the three numbers, which is $\frac{3!}{2!} = 3$. Similarly, there are six ways to order 1 2 3 and only one way to order 2 2 2.
- Therefore, there are $3 + 6 + 1 = 10$ ordered triples that add up to 6.

6 Gold.

- Note that if we switch a with c and b with d, we obtain the exact same expression. It follows that if $a = w$, $b = x$, $c = y$ and $d = z$, the equation is also satisfied if $a = y$, $b = z$, $c = w$, and $d = x$.
- Only one pair of positive integers multiplies to 3: 1 and 3. Without loss of generality, let us set a equal to 1 and c equal to 3. Setting c equivalent to 1 and a equivalent to 3 will yield

the same results except c's value will be switched with a's value and d's value will be switched with b's value.

- We used this method to reduce the number of values that had to be tested when factoring trinomials back in Chapter 1.

- Now we test values of b and d. The four ordered pairs of positive integers that multiply to 6 are 1 and 6, 2 and 3, 3 and 2, 6 and 1. Testing all four, we find one possible answer: $a = 1, b = 2, c = 3, d = 3$.

7 Gold.

- Let us find the number of unordered pairs that satisfy the equation and then find the number of ordered pairs from there.

- Without loss of generality, let us make $x \leq y$. We can do this because every set of x and y in which x is greater than y can be reversed to form a set where $x \leq y$. However, the actual set of two numbers is still the same.

- The minimum possible value of x is 7, as if x gets lower, the term $\frac{1}{y}$ will have to be less than or equal to 0. The maximum possible value of x is 12, as $\frac{1}{12} + \frac{1}{12} = \frac{1}{6}$, and if x gets greater than 12, y will have to be less than x, which violates $x \leq y$.

- We now have to solve 6 quick equations to account for every possibility.

- Setting x equal to 7 yields the equation $\frac{1}{7} + \frac{1}{y} = \frac{1}{6}$. Solving, we find that $y = 42$.

- Similarly,

  setting x equal to 8 yields that $y = 24$,

  setting x equal to 9 yields that $y = 18$,

  setting x equal to 10 yields that $y = 15$,

  setting x equal to 11 does not yield a positive integer solution for y,

  and setting x equal to 12 yields that $y = 12$.

- Therefore, there are 5 sets of positive integers that satisfy the equation. Each one of these sets can form two ordered pairs by switching x and y except for 12 and 12, so the final answer is 9 ordered pairs.

Chapter 3

Number Theory

Part 1: Basic Number Theory

Number theory is, simply put, the theory of numbers. It is where
mathematics began, and today it holds some of the simplest yet most
difficult unsolved problems in the field.

Let us begin with the basics:

Whole numbers are the numbers 0, 1, 2, 3, ... up to positive
infinity.

Natural numbers are defined similarly except 0 is not included.

Integers include whole numbers and their negative correspon-
dents. −1000, 0, and 453 are all integers. Integers are also called
integral numbers.

Number theory is often concerned with factors of positive
integers.

What are factors? A whole number's factors are the positive
integers that divide it evenly. If you divide a number by one of
its factors, you should obtain a whole number. For the purpose of
elementary number theory, consider only the positive factors of
positive integers.

Competitive Math for Middle School: Algebra, Probability, and Number Theory
Vinod Krishnamoorthy
Copyright © 2018 Pan Stanford Publishing Pte. Ltd.
ISBN 978-981-4774-13-0 (Paperback), 978-1-315-19663-3 (eBook)
www.panstanford.com

For example, the factors of 10 are 1, 2, 5, and 10, as 10 is divisible by each of them.

Factors come in pairs. Since 10 is divisible by 2, it has to be divisible by something else that when multiplied by 2 becomes 10. The factor that completes the pair is 5.

In the case of any positive integer n, if another positive integer m is a factor of n, $\frac{n}{m}$ must also be a positive integer and a factor of n; $\frac{n}{m}$ must be a positive integer by the definition of a factor, and $\frac{n}{m}$ is a factor of n because $n \div \frac{n}{m}$ produces a positive integer, m.

For example, 5 is a factor of 119,324,105, so 119,324,105/5 is a positive integer and another factor of 119,324,105.

A positive integer's multiples are the integers that it is a factor of. Since there are infinite positive integers, every integer has an infinite number of multiples.

Prime numbers are positive integers with no factors except one and themselves. Composite numbers are positive integers that are not prime.

An example of a prime number is 2, since its only factors are 1 and itself. However, 4 is composite, because in addition to 1 and itself, it has another factor, 2.

Example 1: Which of the following are prime: 13, 18, and 39?

- 13 has no factors other than 1 and itself, so it is prime.
- 18 has at least one factor because 18/2 yields an integral result. Therefore, 18 is composite.
- Notice that 39 is divisible by 3: $39/3 = 13$. Therefore, 39 is also composite.

Every positive integer greater than 1 has exactly one prime factorization; i.e. a sequence of prime numbers that when multiplied results in the positive integer. Also, no two positive integers have the same prime factorization.

The prime factorization of a number contains a lot of information about the number, and it also serves as a unique identification tag.

The prime factorization of a prime number, for example, 7, is just itself. But for a composite number, such as 42, you must keep "factoring out" its prime factors until you cannot do it anymore.

In the case of 42, we can factor out 2, which yields 2×21, and then we can factor out 3 from the 21, which yields $2 \times 3 \times 7$. All of these are prime, so this is the prime factorization of 42.

Example 2: Find the prime factorization of 12.

- 12 is divisible by 2, so we factor 12 into 2×6.
- 6 is also divisible by 2, so we can further factor 12 into $2 \times 2 \times 3$, or $2^2 \times 3$.
- 2 and 3 are both prime, so we are done.

Example 3: Find the prime factorization of 23.

- To find a number's prime factorization, we have to find its factors and then "factor" them out.
- We find that 23 has no factors besides 1 and itself, so it is prime. Therefore, its prime factorization is just 23.

We can also find prime factorizations by factoring out larger composite numbers and then breaking them down separately.

Example 4: Find the prime factorization of 700.

- $700 = 100 \times 7$. 100's prime factorization is $2 \times 2 \times 5 \times 5$ or $2^2 \times 5^2$, so 700 is equivalent to $2^2 \times 5^2 \times 7$.

A prime factorization in simplest form contains exactly one term per prime factor. For example, $2^2 \times 5 \times 7 \times 5 \times 2$ is not in simplest form, as there are two terms with 2 as their base and two terms with 5 as their base. The simplest form of this would be $2^3 \times 5^2 \times 7$.

Example 5: Put the prime factorization $2 \times 2 \times 3 \times 3 \times 5 \times 7 \times 7$ into simplest form.

- What we must do is combine like terms. $2 \times 2 = 2^2$, $3 \times 3 = 3^2$, and $7 \times 7 = 7^2$.
- Our answer is $2^2 \times 3^2 \times 5 \times 7$.

0 and 1 are numbers that are neither prime nor composite. Therefore, the first prime number is 2. The multiples of 2 are the even numbers. Multiples of 2 greater than 2 cannot be prime, since they have 2 as a factor other than 1 and themselves. Therefore, 2 is the only even prime number.

All factors of a number are subsets of its prime factorization. Therefore, you can tell if a number is divisible by another based on their prime factorizations—it is divisible if the quotient of their prime factorizations yield an integral result. Here it is important to recall the properties of exponents—an expression such as $\dfrac{a^b \times c^d}{a^e \times c^f} = a^{b-e} \times c^{d-f}$.

You can also list out all factors of a number by going through the subsets of its prime factorization.

Example 6: Given that the prime factorization of 1620 is $3^4 \times 2^2 \times 5$, is 1620 divisible by 243?

- We can find that the prime factorization of 243 is 3^5. 3^5 is not a subset of $3^4 \times 2^2 \times 5$, so 1620 is not divisible by 243.

- Here is another way of thinking about this problem: $\dfrac{3^4 \times 2^2 \times 5}{3^5}$ yields $\dfrac{2^2 \times 5}{3}$, using properties of exponents. $2^2 \times 5$ has no power of 3 in its prime factorization, so $2^2 \times 5$ cannot be divisible by 3 and this expression cannot be equivalent to a positive integer, because 3 cannot possibly cancel with a power of 2 or a power of 5.

Example 7: List all factors of 100.

- The prime factorization of 100 is $2^2 \times 5^2$. All subsets of this prime factorization are factors of 100. These are 1, $2^1, 2^2$, 5^1, 5^2, $2^1 \times 5^1$, $2^2 \times 5^1$, $2^1 \times 5^2$ and $2^2 \times 5^2$. If it is easier, you can also list out all of the subsets by writing

- 2^0 × 5^0, 2^1 × 5^0, 2^2 × 5^0,

- 2^0 × 5^1, 2^1 × 5^1, 2^2 × 5^1,

- 2^0 × 5^2, 2^1 × 5^2, 2^2 × 5^2.

- These simplify to 1, 2, 4, 5, 10, 20, 25, 50, and 100.

There are infinitely many prime numbers, so how can we tell whether a number is prime or not besides rigorous testing?

First, know that if a number has no *prime* factors, then it is prime. This is because all composite factors of a number can be written as the product of primes, so you do not have to worry about testing the composite factors.

If any prime below an integer divides it evenly, the integer is not prime. For example, in the case of 323, we find that 323/17 produces a positive integer, so 323 is composite.

Here is another trick: if we are solely trying to determine whether a number is prime or composite, we only have to test divisibility by primes lower than its square root. This is due to the fact that factors come in pairs, so if we test one factor in a pair, testing the other is unnecessary.

In spite of these simple tricks, proving that a large number is prime is a time-consuming task. The largest known prime number is $2^{74,207,281} - 1$, also known as a Mersenne prime because it is one less than a power of 2. This particular prime number has 22,338,618 digits.

Problems: Basic Number Theory

1 Silver. List all factors of 90.

2 Bronze. List the factors of 20.

3 Bronze. Is 25 prime or composite?

4 Bronze. Is 19 prime or composite?

5 Bronze. Find the prime factorization of 34.

6 Silver. Find the prime factorization of 200.

7 Bronze. List the first 12 prime numbers.

8 Gold. Find the prime factorization of 12^{12}.

9 Silver. Is 5200 divisible by 65 × 8?

10 Silver. Find the prime factorization of $\dfrac{2^3 \times 5^6}{2^2 \times 5^3}$.

Part 2: Counting Factors, GCF, LCM

How many factors does a positive integer have?

Find the positive integer's prime factorization and put it in simplest form. Take the exponents of the prime factors, add 1 to each of them, and multiply them. The product equals the number of factors of the original positive integer.

Example 1: Find the number of factors of 120.

- $120 = 2^3 \times 3 \times 5$. $(3+1)(1+1)(1+1) = 16$ factors.

Mathematical Justification: Let us look at 18. The prime factorization of 18 is 2×3^2. For the purpose of this justification, we will write it as $2 \times 3 \times 3$.

- For a positive integer x to be a factor of $2 \times 3 \times 3$, $\dfrac{2 \times 3 \times 3}{x}$ has to produce an integral (an integer) result.

- What are the values of x that work? 2 and 3 work, and so do $2 \times 3, 3 \times 3$, and $2 \times 3 \times 3$. All subsets of $\{2,3,3\}$ are factors of $2 \times 3 \times 3$ when their contents are multiplied.

- How many subsets of 2, 3 and 3 exist? The answer is 6, so there are 6 factors of 18. This reasoning can also be used to show that every factor of a positive integer is a subset of its prime factorization.

If a number is a perfect square, each term in its prime factorization must be a perfect square as well. The same logic goes for perfect cubes, perfect fourth powers, etc.

For example, $2^2 \times 3^4 \times 5^2 \times 7^6$ is a perfect square, as all of the terms in its prime factorization in simplest form are equivalent to something squared: $2^2 = 2^2, 3^4 = (3^2)^2, 5^2 = 5^2$, and $7^6 = (7^3)^2$.

Another way to put it is all perfect squares have their distinct prime factors raised to powers that are divisible by 2 in their prime factorizations. Also, all perfect cubes have their distinct prime factors raised to powers that are divisible by 3 in their prime factorizations, all perfect fourth powers have their distinct prime factors raised to powers that are divisible by 4 in their prime factorizations, etc.

Here is an interesting observation. The only numbers with an odd number of factors are perfect squares, and all perfect squares have an odd number of factors.

Why? All of the exponents of the distinct prime factors in a perfect square's prime factorization are even. For the product of multiple whole numbers to be odd, all of the whole numbers must be odd, since as soon as one multiple of 2 enters the product the whole thing becomes divisible by 2.

One added to an even number yields an odd number, so the only way for the number of factors of a positive integer to be odd is if all of the powers of the distinct prime factors in its prime factorization are even.

A common factor to two numbers is a factor of both numbers. Two numbers do not have a common factor (other than 1) if they do not have a common prime factor.

A common multiple to two numbers is a multiple of both numbers.

Two numbers are *relatively prime* if their greatest common factor is 1. This means that they have no common factors other than 1, and their prime factorizations have no common values.

Due to this definition, 1 is relatively prime to all positive integers.

Two important concepts in number theory are greatest common factor (GCF for short) and least common multiple (LCM for short).

The greatest common factor of two or more positive integers is the greatest positive integer that each of them is divisible by.

The least common multiple of two or more positive integers is the smallest positive integer that is a multiple them all. Greatest common factor (GCF) is sometimes called greatest common divisor (GCD).

Greatest common factors are used to factor expressions and simplify fractions.

Calculating GCF a.k.a. GCD:

- Find the prime factorization in simplest form of each number.

- Find all distinct prime factors among the prime factorizations.

- Find the lowest power of each distinct prime factor (the power could be 0) and multiply the terms containing the powers to form the prime factorization of the GCF.

You are taking the greatest power of each distinct prime factor that divides both numbers evenly and then multiplying them to form the greatest common factor.

Example 2: Find the greatest common factor of 150 and 375:

- The prime factorization of 150 is $2 \times 3 \times 5^2$, and the prime factorization of 375 is 3×5^3.

- The lowest power of 2 in either prime factorization is 2^0 or 1, the term containing the lowest power of 3 is just 3, and the term containing the lowest power of 5 is 5^2. Therefore, the GCF is $1 \times 3 \times 5^2$ or 75.

Recall that for any factor y of a positive integer x, y's prime factorization is a subset of x's prime factorization. The prime factorization of any factor of y will be a subset of y's prime factorization.

However, the factor's prime factorization is also guaranteed to be a subset of x's prime factorization, as a subset of a subset of a set is still a subset of the original set.

Therefore, all factors of a factor of a number are factors of the original number. For example, 50 is a factor of 100, so all factors of 50 are factors of 100.

Knowing the greatest common factor of two numbers, we can find all of their common factors.

Example 3: How many factors do 84 and 18 have in common?

- As in many number theory problems, the first step is to find the prime factorizations of the positive integers. The prime factorization of 84 is $2^2 \times 3 \times 7$, and the prime factorization of 18 is 2×3^2.

- The greatest common factor of the two is 2×3 or 6.

- 6 is the greatest common factor of these two numbers, but it is not their only common factor. Every factor of 6 will also be a common factor. Another way of looking at it is every subset of one 2 and one 3 will divide both 84 and 18 evenly.

- From here we find that 84 and 18 have four common factors.

Calculating LCM:

- Find the prime factorization in simplest form of each number.

- Find all distinct prime factors among the prime factorizations.

- Find the highest power of each distinct prime factor and multiply the terms containing the powers to form the prime factorization of the LCM.

You are trying to find the least positive integer that both numbers are a factor of, and this can be done on the level of individual prime factors.

Example 4: Find the least common multiple of 35 and 100.

- The prime factorization of 35 is 5×7, and the prime factorization of 100 is $2^2 \times 5^2$.

- The term containing the highest power of 2 is 2^2, the term containing the highest power of 5 is 5^2, and the term containing the highest power of 7 is just 7. Therefore, the LCM is $2^2 \times 5^2 \times 7$, or 700.

The least common multiple of two numbers multiplied by their greatest common factor is equivalent to the product of the two numbers. This is because for any distinct prime factor in the prime factorizations of the numbers, one of the prime factorizations contains its highest power and the other prime factorization contains its lowest power.

Problems: Counting Factors, GCF, LCM

1 Silver. a and b are positive integers. If $a \times b = 72$, how many distinct values can a take on?

2 Silver. Find the greatest common factor of 14 and 77.

3 Silver. Find the greatest common factor of 660 and 108.

4 Bronze. What is the greatest common factor of 125 and 250?

5 Silver. Find the greatest common factor of 360, 900, and 84.

6 Gold. $(x + y)$ and $(2x + 3y)$ are integers. Find all ordered pairs (xy) that satisfy $(x + y)(2x + 3y) = 13$.

7 Silver. Find the greatest common factor of 10! and 13!.

8 Bronze. Find the least common multiple of 12 and 16.

9 Silver. Find the least common multiple of 60, 72, and 40.

10 Silver. Find the prime factorization of the least common multiple of $11^{12} \times 14^3$ and $21^2 \times 2^9$.

11 Gold. Two positive integers have a greatest common factor of 25, a least common multiple of 825, and a sum of 350. Find the two positive integers.

12 Gold. List all common factors of 95 and 950.

13 Silver. Is $6^{101} \times 13^5 \times 34^4$ relatively prime to $32^{38} \times 19^5$?

14 Silver. Are 33^{11} and 16^{22} relatively prime?

15 Silver. Write 495 as the product of three relatively prime positive integers greater than 1.

16 Silver. List all subsets of the set $\{1,2,3,4\}$ in which there are two or more elements and all elements are relatively prime.

Part 3: Bases

So far, we have only dealt with base 10 numbers. We use a base 10 number system, meaning that we have 10 different digits. However, it is possible to have number systems of any other whole number base.

To examine other number systems, we first must examine our own. Let us review the concept of place value.

For any whole number, the ones digit adds one times its value to the number. For example, the value added to the number 25 by the digit 5 is 5.

The tens digit adds 10 times its value to the number. The value added to the number 25 by the 2 is 20, as it is the tens digit of this number.

In the number 1349, the 1 holds a value of 1000, the 3 holds a value of 300, the 4 holds a value of 40, and the 9 holds a value of 9. $1000 + 300 + 40 + 9$ adds back to 1349.

This principle continues onwards with increasing powers of 10 and even extends the other way into the decimal range.

Let us examine the number 15. In terms of the number 10, $15 = (10^0 \times 5) + (10^1 \times 1)$. Let us also examine the number 356. 356 can be written as $(10^0 \times 6) + (10^1 \times 5) + (10^2 \times 3)$. Do you see the pattern now?

This also holds true for every other base as well. A value x base y is denoted as x_y.

$10_2 = (2^0 \times 0) + (2^1 \times 1) = 2$ in base 10.

Similarly, $12_{16} = (16^0 \times 2) + (16^1 \times 1) = 18_{10}$.

One more: $120123_4 = (4^0 \times 3) + (4^1 \times 2) + (4^2 \times 1) + (4^3 \times 0) + (4^4 \times 2) + (4^5 \times 1) = 3 + 8 + 16 + 0 + 512 + 1024 = 1563_{10}$

When dealing with bases, do not forget that there are only n possible digits in a base n number. Therefore, 21_2 is not a real number, because base 2 only contains the digits 0 and 1. Similarly, 743_5 is not a real number because base 5 numbers can only have the digits 0, 1, 2, 3, and 4.

You may also have noticed that you cannot fully represent bases higher than 10 without extra digits. We need a digit that is equivalent to 10 in order to complete base 11. To do this, we use capital letters to represent the digits higher than 9, starting with A as 10 and continuing onward.

For example,

$$AB_{13} = \left(13^0 \times 11\right) + \left(13^1 \times 10\right) = 11 + 130 = 141_{10}.$$

Also, $2F1_{16} = \left(16^0 \times 1\right) + \left(16^1 \times 15\right) + \left(16^2 \times 2\right) = 753_{10}$

Now that we know how to convert from other bases to base 10, how do we convert from base 10 to other bases?

To do this, we take the number we are converting and continually divide by our desired base. The remainder each time is the digit (start at the units digit and keep adding digits to the left), and the quotient portion is the number to be divided next.

Example 1: Convert 312 to base 6.

- First, we divide 312 by 6. The result is 52 remainder 0, so the units digit is 0. Next we divide 52 by 6. This yields 8 remainder 4, so the tens digit is 4. Dividing 8 by 6 yields 1 remainder 2, so the hundreds digit is 2.

- Dividing 1 by 6, we find that the thousands digit is 1. The quotient portion of $1/6$ is 0. If we keep dividing 0 by 6, we will just get more and more 0's. This will not change the value of the number. Therefore, $312 = 1240_6$.

Problems: Bases

1 Silver. A two-digit positive integer is equivalent to three times the sum of its digits. Find the product of its digits.

2 Silver. Convert 1010_2 to base 10.

3 Silver. Convert 2331_5 to base 10.

4 Silver. Convert $3AB3_{13}$ to base 10.

5 Silver. Convert 135_{10} to base 4.

6 Silver. Convert 3907_{10} to hexadecimal (base 16).

7 Silver. Convert 543_7 to base 9.

8 Silver. Convert 212101_3 to base 12.

9 Silver. Calculate the value of $10102_2 + 11101_2$ in base 2.

10 Silver. Calculate the value of $233_4 \times 122_3$ in base 7.

11 Bronze. The units digit of a two-digit positive integer is A, and the tens digit of the positive integer is B. Find the value of the positive integer in terms of A and B.

12 Gold. When the digits of a three-digit positive integer are reversed, the result is equivalent to 198 less than the original positive integer. What is the difference between the hundreds digit and the ones digit of the original integer?

Part 4: Modular Arithmetic

Modular arithmetic is arithmetic involving remainders. x modulo y means the remainder when x is divided by y. Entire equations can be written in modulo or mod for short. In a modular equation, the equal sign is replaced with $\equiv$ (said as "congruent").

If we have that $a \equiv b$ mod x, a and b have the same remainder when divided by x. For example, $3 \equiv 8$ mod 5, $2 \equiv 4$ mod 2, and $1 \equiv 101$ mod 100.

Negative integers also count. For example, $-1 \equiv (-1 + 4)$ or 3 in mod 4, and $-100 \equiv 2$ (mod 3). For the purpose of modular arithmetic, it is necessary to consider the negative multiples of positive integers. The negative multiples of 3 are $-3, -6, -9, -12$, etc.

Here is an interpretation of modular congruence explained using an example:

If x is congruent to 3 in mod 4, x is also congruent to $-5, -1, 7$, 11, 15, 19, etc. (any $3 + 4n$ where n is an integer) because they all produce a remainder of 3 when divided by 4.

Another way of putting it is if x is congruent to 3 mod 4, x is 3 greater than a multiple of 4.

Example 1: Solve $x \equiv 32 \pmod{12}$.

- Since 32 itself is congruent to 8 in mod 12 ($32/12 = 2$ remainder 8), we can replace 32 with 8, leaving us with $x \equiv 8 \pmod{12}$.

- This is in simplest form, as one cannot replace 8 with anything simpler in mod 12.

In mod n, there are n distinct non-congruent values. For example, in mod 4, there are only four distinct values that are not congruent to each other: 0, 1, 2, and 3. Every other whole number in mod 4 is congruent to one of these four values.

In these equations, we can combine terms through addition, subtraction, multiplication, and sometimes division.

1. $(a \bmod x) + (b \bmod x) = (a + b) \bmod x$
2. $(a \bmod x) - (b \bmod x) = (a - b) \bmod x$,
3. $(a \bmod x) \times (b \bmod x) = (a \times b) \bmod x$
4. $(a \bmod x)/(b \bmod x) = (a/b) \bmod x$. If b and x are relatively prime.

With these properties, we can balance modular equations. Note that fractions are invalid answers in modular arithmetic, as when two positive integers are divided the remainder cannot be a fraction.

Example 2: Solve $x + 3 \equiv 19 \bmod 3$.

- By these properties, we are able to subtract 3 from both sides, which leaves us with $x \equiv 16 \bmod 3$.

- The remainder when 16 is divided by 3 is 1, so $16 \equiv 1 \bmod 3$ and therefore we can simplify our equation to $x \equiv 1 \bmod 3$. Although this is the solution to the equation, there are many possible values for x. x can be $-8, -2, 1, 4, 13$, etc.

Note that if we had received a modular equation that gave us a fractional answer for x, such as 1.5 or $\frac{3}{2}$, the answer would be invalid, as modular arithmetic only concerns whole number remainders.

So how do we solve an equation such as $5x \equiv 2 \pmod 7$?

- Since $5x \equiv 2$ in mod 7, $5x$ is also congruent to $9, 16, 23, 30$, etc. in mod 7 as well.

- What is the first number that is divisible by 5 in this list? The answer is 30. The equation $5x \equiv 30 \pmod 7$ yields the result $x \equiv 6 \pmod 7$, our final answer. Since 5 is relatively prime to 7, our division in modular arithmetic is valid.

Example 3: $8x + 4 \equiv 12 \pmod 8$.

- Subtracting 4 from both sides yields $8x \equiv 8 \pmod 8$. However, we cannot divide both sides by 8 in this case, as $8 \equiv 0 \pmod 8$, and dividing by 0 is prohibited.

- We must notice that $8x \equiv 8$ will always be true no matter what the value of x, as 8 multiplied by any number will always be congruent to 0 in mod 8.

- Therefore, all integers satisfy the modular equation.

Example 4: If the remainder when x is divided by 9 is 5, what is the remainder when $4x$ is divided by 9?

- We have that $x \equiv 5 \pmod 9$. We are trying to find what $4x$ is congruent to in mod 9.

- To do this, we multiply both sides of this equation by 4. This leaves us with $4x \equiv 20 \pmod 9$. 20 is congruent to 2 in mod 9, so the remainder when $4x$ is divided by 9 is 2.

Note that if you take the value of a base 10 positive integer mod 10, you obtain its units digit, if you take the value mod 100, you obtain its last two digits, and so forth. This fact can be used to solve many problems.

Problems: Modular Arithmetic

1 Bronze. Solve $x + 2 \equiv 6 \pmod 5$.

2 Bronze. Solve $x + 7 \equiv 4 \pmod 5$.

3 Bronze. Solve $x + 4 \equiv 88 \pmod 6$.

4 Bronze. $x + 4 \equiv 9 \pmod{12}$. Solve for x.

5 Silver. Is the sum of 42 even numbers and 23 odd numbers even or odd? What about the sum of 299 even numbers and 314 odd numbers?

6 Silver. When a certain positive integer is divided by 17, the remainder is 11. The positive integer is then multiplied by 3 and then added to 11. What is the remainder when this result is divided by 17?

7 Silver. $15x + 3 \equiv 22 + 2x \pmod 7$. Solve for x.

8 Silver. What is the 9010th term of the repeating sequence 1, 2, 3, 1, 2, 3, 1, 2, 3, 1, 2, 3, ...?

9 Gold. What is the units digit of 4929^{732}?

10 Gold. What is the units digit of 22^{401}?

11 Silver. January 1 of a particular year lands on a Wednesday. The year is not a leap year, and neither is the following year. What day of the week is January 1 of the following year?

12 Gold. It is now February 17, 1973. If November 29, 1972, was a Wednesday, what day of the week is March 30, 1974?

13 Gold. Find the tens digit of 103^{10}.

14 Silver. If $x \equiv 7 \pmod{15}$, which of the following is true?

$$x \equiv 33 \pmod{15}$$
$$x \equiv 1589 \pmod{15}.$$
$$x \equiv 2 \pmod 5.$$

15 Platinum. For any base 8 number, dividing by 4^n will yield its last three digits as a remainder. Find n.

Part 5: Divisibility Tricks

Divisibility tricks help test if a given integer is divisible by another.

2: If the number is even, it is divisible by 2.

3: If the sum of the digits of a number is divisible by 3, then the number is divisible by 3.

Proof:

- Let us say we are testing the number $abcd$, where a, b, c, and d are digits, not variables.

- We first write the number being tested as $1000a + 100b + 10c + d$. For this number to be divisible by 3, it has to be congruent to 0 in mod 3.

- Notice that all powers of 10 are congruent to 1 in mod 3. Therefore, $1000a + 100b + 10c + d$ (mod 3) can be simplified into $(1)\,a + (1)\,b + (1)\,c + d$ (mod 3), the sum of the number's digits.

- The same process works no matter how long the original number is.

4: If the last two digits of a number are divisible by 4, then the number is divisible by 4.

Proof:

- The remainder when a number x is divided by 4 is the value of x in mod 4. x is equivalent to $10^y \times a + 10^{y-1} \times b + 10^{y-2} \times c + \cdots$, where 10^y is the place value of x's first digit and a, b, c, ... are the digits of x.

- Note that all powers of 10 except for 10^0 and 10^1 are divisible by 4, since they have 2^2 or greater in their prime factorizations. Therefore, all terms in the expansion $(10^y \times a, 10^{y-1} \times b,$ etc.$)$, cancel out to become 0 in mod 4, leaving just the last two digits.

5: If the number ends in a 5 or a 0, then the number is divisible by 5.

6: If a number is divisible by 2 and 3, then it is divisible by 6.

7: There are numerous divisibility tricks for 7, but most of them involve a lot of calculations. Test for divisibility by 7 directly.

8: If the last three digits of a number are divisible by 8, then it is divisible by 8.

  Proof: This proof is quite similar to the proof of 4's divisibility rule.

9: If the sum of the digits of a number is divisible by 9, then the number is divisible by 9.

  Proof: This proof is quite similar to the proof of 3's divisibility rule.

10: If a number ends in a 0, then it is divisible by 10.

11: Sum up the first, third, fifth, etc. digits, and then sum up the second, fourth, sixth, etc. digits. If the difference of these sums is 0 or a multiple of 11, the number is a multiple of 11 as well.

Other composite numbers: To find whether a number is divisible by n, split n into two or more relatively prime factors and test for divisibility with each of them. If the number is divisible by all of the relatively prime factors of n, it is divisible by n. For example, for a number to be divisible by 12, it has to be divisible by 3 and 4, and for a number to be divisible by 440, it has to be divisible by 5, 8, and 11. However, a number is not necessarily divisible by 16 if it is divisible by 2 and 8, since 2 and 8 are not relatively prime.

Problems: Divisibility Tricks

1 Silver. Is 161 prime or composite?

2 Silver. Is 1001 prime or composite?

3 Gold. Is 2011 prime or composite?

4 Bronze. Is 1009 divisible by 4?

5 Bronze. Is 11, 301 divisible by 5?

6 Silver. Is 97,884 divisible by 6?

7 Silver. Is 10,343,662,626,934,841,435 divisible by 99?

8 Silver. Is $111, 632, 434, 434, 430$ divisible by 15?

Solutions Manual

Part 1: Basic Number Theory

1 Silver.

- We first take the prime factorization of 90. $90 = 3 \times 30 = 3 \times 3 \times 10 = 3 \times 3 \times 2 \times 5 = 2 \times 3^2 \times 5$. The subsets of the prime factorization are the factors of 90. Let us list them with a noticeable pattern:
  - $2^0 \times 3^0 \times 5^0, 2^0 \times 3^1 \times 5^0, 2^0 \times 3^2 \times 5^0, 2^0 \times 3^0 \times 5^1, 2^0 \times 3^1 \times 5^1, 2^0 \times 3^2 \times 5^1,$
  - $2^1 \times 3^0 \times 5^0, 2^1 \times 3^1 \times 5^0, 2^1 \times 3^2 \times 5^0, 2^1 \times 3^0 \times 5^1, 2^1 \times 3^1 \times 5^1, 2^1 \times 3^2 \times 5^1.$
- Simplifying, we find that the factors of 90 are 1, 3, 9, 5, 15, 45, 2, 6, 18, 10, 30, and 90. This answer is perfectly fine, but if you want to list them in order, you would write 1, 2, 3, 6, 9, 10, 15, 18, 30, 45, 60, 90.

2 Bronze.

- The positive integers, in order, that divide 20 evenly are 1, 2, 4, 5, 10, and 20.

3 Bronze.

- 25 has a factor other than 1 and itself: 5. Therefore, it is composite.

4 Bronze.

- Through inspection and perhaps a few tests, we realize that 19 has no factors other than 1 and itself. Therefore, it is prime.

5 Bronze.

- 34 is divisible by 2, so we can factor it into 2×17. 2 and 17 are both prime, so we are done.

6 Silver.

- 200 is divisible by both 2 and 5, so it can be factored into $2 \times 5 \times 20$. 20 is divisible by both 2 and 5 as well, so 200 can be further factored into $2 \times 5 \times 2 \times 5 \times 2$, or $2^3 \times 5^2$.

7 Bronze.

- The first 12 primes in order are: 2, 3, 5, 7, 11, 13, 17, 19, 23, 29, 31, 37.

8 Gold.

- How do we write this as a product of primes? $12 = 2^2 \times 3$. Therefore, $12^{12} = (2^2 \times 3)^{12}$, which can be expanded into $(2^2)^{12} \times 3^{12}$.

- Simplifying yields $2^{24} \times 3^{12}$, which is the final prime factorization.

9 Silver.

- Let us find the prime factorization of 5200.

- $5200 = 52 \times 100$.

- $52 = 13 \times 4 = 2^2 \times 13$ and $100 = 2 \times 50 = 2 \times 2 \times 25 = 2^2 \times 5^2$. Therefore, $5200 = 2^2 \times 13 \times 2^2 \times 5^2 = 2^4 \times 5^2 \times 13$.

- Now we find the prime factorization of 65×8. $65 = 5 \times 13$, and $8 = 2^3$. Therefore, the prime factorization of 65×8 is $2^3 \times 5 \times 13$.

- For 5200 to be divisible by 65×8, the prime factorization of 65×8 has to be a subset of the prime factorization of 5200. $2^3 \times 5 \times 13$ is a subset of $2^4 \times 5^2 \times 13$, so 5200 is divisible by 65×8.

10 Silver.

- Recall that $x^{y-z} = \dfrac{x^y}{x^z}$. Applying this principle, we simplify the given expression to $2^1 \times 5^3$. We are asked for the prime factorization of the expression, not the expression's value, so we are done.

Part 2: Counting Factors, GCF, LCM

1 Silver.

- Dividing both sides of the equation by a, we find that $b = \dfrac{72}{a}$. For b to be a positive integer, $\dfrac{72}{a}$ must produce an integral result. Therefore, a must be a factor of 72. Also, each value of a corresponds to a distinct solution.

- How many factors does 72 have? 72's prime factorization is $2^3 \times 3^2$. Therefore, it has $(3 + 1)(2 + 1)$ or 12 factors.

2 Silver.

- The prime factorization of 14 is 7×2, and the prime factorization of 77 is 7×11. The largest factor that is common to 7×2 and 7×11 is 7.

3 Silver.

- The prime factorization of 660 is $2^2 \times 3 \times 5 \times 11$, and the prime factorization of 108 is $2^2 \times 3^3$. The four distinct prime factors between the two prime factorizations are 2, 3, 5, and 11.

- The term containing the lowest power of 2 is 2^2, the term containing the lowest power of 3 is 3^1, and the lowest powers of 5 and 11 in either prime factorization are 5^0 and 11^0 respectively. $2^2 \times 3 \times 5^0 \times 11^0 = 12$, our answer.

4 Bronze.

- 125 divides 250 evenly: $250/125 = 2$. This means that 125 is a common factor to 125 and 250. There cannot be a greater common factor, as nothing larger than 125 can be a factor of 125.

- Therefore, 125 is the greatest common factor.

5 Silver.

- 360: $2^3 \times 3^2 \times 5$, 900: $2^2 \times 3^2 \times 5^2$, 84:$2^2 \times 3 \times 7$
- The distinct prime factors are 2, 3, 5, and 7.
- The term containing the lowest power of 2 is 2^2, the term containing the lowest power of 3 is 3^1, the lowest power of 5 among the prime factorizations is 5^0 (meaning the prime factorization has no 5), and the lowest power of 7 among the prime factorizations is 7^0.
- $2^2 \times 3^1 \times 5^0 \times 7^0 = 4 \times 3 \times 1 \times 1 = 12$.

6 Gold.

- $(x + y)$ and $(2x + 3y)$ are integers that multiply to 13. How many pairs of integers multiply to 13?
- 13 is prime, so the only pair of positive integers that multiply to 13 is 1 and 13. However, integers can be negative. We must also consider the pair -1 and -13.
- There are 4 ways that we can assign the values to the two expressions. We can assign 1 to $(x + y)$ and 13 to $(2x + 3y)$, which yields the system of equations $x + y = 1$ and $2x + 3y = 13$. This system yields the solution $(-10, 11)$.
- We can also assign the 13 to $(x + y)$ and 1 to $(2x + 3y)$, which yields the system of equations $x + y = 13$ and $2x + 3y = 1$. This system yields the solution $(38, -25)$.
- Next, we set $(x + y)$ equal to -1 and $(2x + 3y)$ equal to -13. This yields the system of equations $x + y = -1$ and $2x + 3y = -13$, which yields the solution $(10, -11)$.
- Lastly, we set $(x + y)$ equal to -13 and $(2x + 3y)$ equal to -1. This yields the system of equations $x + y = -13$ and $2x + 3y = -1$, which yields the solution $(-38, 25)$.

7 Silver.

- What is the greatest positive integer that divides both 10! and 13! evenly? $\dfrac{13!}{10!}$ yields an integral result. Therefore, 10! is a factor of 13!. No factor of 10! can be greater than 10!, so the greatest common factor of 10! and 13! is 10!.

8 Bronze.

- The prime factorization of 12 is $2^2 \times 3$, and the prime factorization of 16 is 2^4. The two distinct prime factors between the prime factorizations are 2 and 3. The term containing the greatest power of 2 is 2^4, and the term containing the greatest power of 3 is 3^1.
- $2^4 \times 3^1 = 48$.

9 Silver.

- The prime factorization of 60 is $2^2 \times 3 \times 5$, the prime factorization of 72 is $2^3 \times 3^2$, and the prime factorization of 40 is $2^3 \times 5$.
- The distinct prime factors among the prime factorizations are 2, 3, and 5. The term containing the greatest power of 5 is 5^1, the term containing the greatest power of 3 is 3^2, and the term containing the greatest power of 2 is 2^3.
- $2^3 \times 3^2 \times 5 = 360$.

10 Silver.

- What is the prime factorization of $11^{12} \times 14^3$? 11 is already prime, so we do not have to do anything there.
- However, 14 is composite. Therefore, we must split 14^3 up into the product of prime factors. $14^3 = (7 \times 2)^3 = 7^3 \times 2^3$ by the properties of exponents. Similarly, $21^2 = (3 \times 7)^2 = 3^2 \times 7^2$.
- The two prime factorizations are $2^3 \times 7^3 \times 11^{12}$ and $2^9 \times 3^2 \times 7^2$. The distinct prime factors are 2, 3, 7, and 11. The term containing the greatest power of 2 is 2^9, the term containing the greatest power of 3 is 3^2, the term containing the greatest power of 7 is 7^3, and the term containing the greatest power of 11 is 11^{12}.
- Therefore, the prime factorization of the answer is $2^9 \times 3^2 \times 7^3 \times 11^{12}$.

11 Gold.

- 25, the greatest common factor of the two numbers, equals 5^2. Therefore, the term containing the lowest power of 5 in either of the two numbers' prime factorizations must be 5^2, and there must be no other common factor to the two.

- 825, the least common multiple of the two numbers, equals $3 \times 5^2 \times 11$. Therefore, the highest power of 5 in either of the two numbers' prime factorizations must be 5^2, the highest power of 3 must be 3^1, and the highest power of 11 must be 11^1.

- We can now conclude that both of the numbers must have 5^2 in their prime factorization, as the highest and lowest powers of 5 in either of the prime factorizations are both 5^2.

- We can also conclude that exactly one of the numbers has 3^1 in its prime factorization (if both had a 3^1, 3^1 would be in the greatest common factor) and exactly one of the numbers has a 11^1 in its prime factorization.

- Going through all of the possible cases, the two numbers can be either 5^2 and $3 \times 5^2 \times 11$ or 3×5^2 and $5^2 \times 11$. The latter pair has a sum of 350, so the two numbers are 75 and 275.

12 Gold.

- Notice that $950/95 = 10$. Therefore, 95 is the greatest common factor of the two numbers.

- What is the prime factorization of 95? $95 = 5 \times 19$. Every subset of one 5 and one 19 will be a common factor. These four subsets are 1 ($5^0 \times 19^0$), 5 ($5^1 \times 19^0$), 19 ($5^0 \times 19^1$), and 95 ($5^1 \times 19^1$).

13 Silver.

- The numbers have 2 as common factor. (They have many, many more common factors as well). Therefore, they are not relatively prime.

14 Silver.

- $33^{11} = 3^{11} \times 11^{11}$, and $16^{22} = (2^4)^{22} = 2^{88}$. These prime factorizations do not have any factors in common, so 33^{11} and 16^{22} are relatively prime.

15 Silver.

- We do not know the three positive integers right away, so we take the prime factorization of 495 first. $495 = 3^2 \times 5 \times 11$.
- To split 495 into the product of relatively prime positive integers, we must split the prime factorization into relatively prime pieces. If the two 3's are separated, two or more of the relatively prime pieces will be divisible by 3, so the three positive integers will not be relatively prime.
- Therefore, one of the positive integers has to be 3^2 or 9. It is easy to see that the other two integers are 5 and 11.

16 Silver.

- Start with the two-element subsets. Those that work are $\{1,2\}$, $\{1,3\}$, $\{1,4\}$, $\{2,3\}$, and $\{3,4\}$, as the greatest common factor of the elements in these sets is 1.
- The only three-element subsets that work are $\{1,2,3\}$ and $\{1,3,4\}$. In $\{1,2,4\}$ and $\{2,3,4\}$, 2 and 4 are not relatively prime.
- The one four-element subset, $\{1,2,3,4\}$, does not work, as 2 and 4 have a common factor of 2.

Part 3: Bases

1 Silver.

- Let us call the tens digit of the number x and the units digit y. The number equals $10x + y$, and the sum of its digits is simply $x + y$. We have that $10x + y = 3(x + y)$. Expanding yields $10x + y = 3x + 3y$.

- In multi-variable problems, it is often helpful to solve for one variable in terms of the other(s). Let us solve for x in terms of y. Doing so, we find that $x = \frac{2}{7}y$.

- How do we solve this? We use the fact that both x and y have to be one-digit whole numbers, since they are digits. For $\frac{2}{7}y$ to be a whole number, y has to equal either 0 or 7 (remember that y cannot be greater than 9).

- 00 is not a two-digit number, so y has to be 7. It follows that $x = (2/7) \times 7 = 2$. Therefore, the number we are looking for is 27. The product of 2 and 7 is 14.

2 Silver.

- 1010_2 can be written as $(2^0 \times 0) + (2^1 \times 1) + (2^2 \times 0) + (2^3 \times 1)$. This equals $0 + 2 + 0 + 8$, or 10.

3 Silver.

- 2331_5 can be written as $(5^0 \times 1) + (5^1 \times 3) + (5^2 \times 3) + (5^3 \times 2)$. This equals $1 + 15 + 75 + 250$, or 341.

4 Silver.

- The digit A has a value of 10 and the digit B has a value of 11. $3AB3_{13}$ can be written as $(13^0 \times 3) + (13^1 \times 11) + (13^2 \times 10) + (13^3 \times 3)$. This equals $3 + 143 + 1690 + 6591$, or 8427.

5 Silver.

- To convert the base 10 number 135 to base 4, we repeatedly divide 135 by 4 and use the remainders as the digits of our result.

- $135/4 = 33$ remainder 3, so the units digit is 3. 33 divided by 4 yields 8 remainder 1, so the tens digit is 1. $8/4 = 2$ remainder 0, so the hundreds digit is 0. Lastly, we divide $2/4$, finding that the thousands digit is 2.

- Putting these together yields 2013_4.

6 Silver.

- Dividing 3907 by 16, we receive a remainder of 3 and a quotient of 244. Dividing 244 by 16, we receive a remainder of 4 and a quotient of 15. Dividing 15 by 16, we receive a remainder of 15 and a quotient of 0.

- Putting the digits together yields $F43_{16}$, as the digit F represents 15.

7 Silver.

- First, we convert the base 7 number to base 10, and then we convert the base 10 number to base 9.
- 543_7 can be written as $(7^0 \times 3) + (7^1 \times 4) + (7^2 \times 5)$ or 276.
- Now, we repeatedly divide 276 by 9. The first result is 30 remainder 6, so the units digit of the base 9 number is 6. Next, we divide 30 by 9, obtaining 3 remainder 3. Therefore, the tens digit is 3. Lastly, we divide 3 by 9, obtaining 0 remainder 3.
- Putting them together yields 336_9.

8 Silver.

- First, we convert the base 3 number to base 10, and then we convert the base 10 number to base 12.
- 212101_3 can be expanded into $(3^0 \times 1) + (3^1 \times 0) + (3^2 \times 1) + (3^3 \times 2) + (3^4 \times 1) + (3^5 \times 2)$. This equals $1 + 0 + 9 + 54 + 81 + 486$, or 631.
- Next, we convert this to base 12. 631/12 equals 52 remainder 7, so the units digit is 7. 52/12 equals 4 remainder 4, so the tens digit is 4. Finally, 4/12 equals 0 remainder 4, so the hundreds digit is 4 as well.
- Putting the digits together yields 447_{12}.

9 Silver.

- Our strategy will be to convert both of the numbers to base 10, find their sum, and then convert the result back to base 2.
- $1010_2 = (2^0 \times 0) + (2^1 \times 1) + (2^2 \times 0) + (2^3 \times 1)$, or 10. $11101_2 = (2^0 \times 1) + (2^1 \times 0) + (2^2 \times 1) + (2^3 \times 1) + (2^4 \times 1)$, or 29. $10 + 29$ equals 39.
- Now for converting 39 to base 2. 39/2 equals 19 remainder 1, so the units digit is 1. 19/2 = 9 remainder 1, so the next digit

to the left is 1. $9/2 = 4$ remainder 1, $4/2 = 2$ remainder 0, $2/2 = 1$ remainder 0, and $1/2$ equals 0 remainder 1. Putting the remainders together, we obtain our answer, 100111_2.

10 Silver.

- To solve the problem, we convert both numbers to base 10, multiply them, and convert the result to base 7.

- $233_4 = (4^0 \times 3) + (4^1 \times 3) + (4^2 \times 2) = 3 + 12 + 32 = 47$. $122_3 = (3^0 \times 2) + (3^1 \times 2) + (3^2 \times 1) = 17$. $47 \times 17 = 799$.

- Now for converting 799 to base 7. $799/7 = 144$ remainder 1, $114/7 = 16$ remainder 2, $16/7 = 2$ remainder 2, and $2/7 = 0$ remainder 2. Putting the remainders together, we obtain the answer, 2221_7.

11 Bronze.

- The digit B is in the tens place, so it adds ten times its value to the overall number. The digit A is in the ones place, so it adds 1 times its value to the overall number. Therefore, the number's value is $10B + A$.

12 Gold.

- Let us call the original three-digit positive integer XYZ, where X is the hundreds digit, Y is the units digit, and Z is the ones digit. The integer is equivalent to $100X + 10Y + Z$ when written in terms of the values of its digits.

- Reversing the digits of XYZ produces the three-digit positive integer ZYX. ZYX is equivalent to $100Z + 10Y + X$. We know from the problem that $100X + 10Y + Z - 198 = 100Z + 10Y + X$.

- Subtracting $10Y$ from both sides yields $100X + Z - 198 = 100Z + X$. We are looking to find the value of $X - Z$. Subtracting $100Z$ and then X from both sides yields $99X - 99Z - 198 = 0$, and adding 198 to both sides yields $99X - 99Z = 198$.

- $99X - 99Z$ can be factored into $99(X - Z)$, leaving us with the equation $99(X - Z) = 198$. Dividing both sides by 99, we find that $X - Z = 2$.

Part 4: Modular Arithmetic

1 Bronze.
- Subtracting 2 from both sides, we find that $x \equiv 4 \pmod 5$.

2 Bronze.
- Subtracting 7 from both sides, we find that $x \equiv -3 \pmod 5$. Since -3 is congruent to 2 in mod 5, the simplest form of the solution to the equation is $x \equiv 2 \pmod 5$.

3 Bronze.
- Subtracting 4 from both sides, we find that $x \equiv 84 \pmod 6$. $84/6$ yields a remainder of 0, so 84 is congruent to 0. The simplest solution to the equation is $x \equiv 0 \pmod 6$.

4 Bronze.
- We subtract 4 from both sides of the equation, obtaining $x \equiv 5 \pmod{12}$.

5 Silver.
- Whether a number is even or odd is depends on its value in mod 2. If the number is congruent to 0 in mod 2, it is even, but if it is congruent to 1, it is odd.
- The sum of 42 even numbers will always be even. Every even number can be replaced with 0 when in mod 2, and the sum of any number of 0's is 0.
- Every odd number can be replaced with 1 in mod 2. The sum of 23 odd numbers in mod 2 is congruent to $1(23)$ mod 2, and $23 \equiv 1$ mod 2. Hence, the first sum is odd.
- From all of this, it follows that the sum of even numbers is always congruent to 0 in mod 2, the sum of an odd number

of odd numbers is congruent to 1 in mod 2, and the sum of an even number of odd numbers is congruent to 0 in mod 2.

- By this logic, the second sum is even.

6 Silver.

- Let us call the number n. When n is divided by 17, the remainder is 11, so $n \equiv 11 \pmod{17}$.

- If n is tripled and then added to 11, $11 + 3n$ is the result. How do we obtain $11 + 3n$ from $n \equiv 11$? We first multiply both sides by 3, which yields $3n \equiv 33 \pmod{17}$, and then we add 11 to both sides, obtaining $3n + 11 \equiv 44 \pmod{17}$.

- 44/17 yields a remainder of 10, so the equation can be simplified to $3n + 11 \equiv 10 \pmod{17}$. Therefore, the desired remainder is 10.

7 Silver.

- Adding and subtracting terms from both sides, we simplify the equation to $13x \equiv 19 \pmod{7}$.

- 13 is congruent to 6 in mod 7 and 19 is congruent to 5 in mod 7, so we can further simplify the equation into $6x \equiv 5 \pmod{7}$. $x \equiv 5/6$ is not a valid answer, but replacing 5 with 12 yields that $x \equiv 2 \pmod{7}$. The division is valid because 6 and 7 are relatively prime.

8 Silver.

- Let us analyze this sequence. The 1st, 4th, 7th, 10th, 13th, ... terms are 1, the 2nd, 5th, 8th, 11th, 14th, ... terms are 2, and the 3rd, 6th, 9th, 12th, 15th, ... terms are 3.

- 1, 4, 7, 10, and 13 all are one more than a multiple of 3, so they are all congruent to 1 in mod 3. Similarly, the terms equivalent to 2 are congruent to 2 in mod 3, and the terms equivalent to 3 are congruent to 0 in mod 3.

- What is 9010 in mod 3? $\dfrac{9010}{3}$ equals 3003 remainder 1, so the 9010th term of this series is 1.

9 Gold.

- We definitely are not going to compute this value and then see what its units digit is. Instead, we must find the value of the expression in mod 10. This will equal the units digit of the expression.

- 4929 (mod 10) $\equiv$ 9 (mod 10), so 4929^{732} (mod 10) $\equiv$ 9^{732} (mod 10).

- The units digit of 9^1 is 9, the units digit of 9^2 is 1, the units digit of 9^3 is $9^2 \times 9$ (mod 10) $\equiv$ 1×9 (mod 10) which is congruent to 9, the units digit of 9^4 is $9^3 \times 9$ (mod 10) $\equiv$ 9×9 (mod 10) which is congruent to 1, and this pattern – 9, 1, 9, 1, 9, 1 . . . – continues.

- Every even power of 9 has units digit 1, and every odd power of 9 has units digit 9, as in the pattern. Therefore, 4929^{732} has units digit 1.

10 Gold.

- Just as we did in our previous units digit problem, we will look for a pattern among the units digits of the powers of 2.

- 2^1 (mod 10) $\equiv$ 2, 2^2 (mod 10) $\equiv$ 4, 2^3 (mod 10) $\equiv$ 8, 2^4 (mod 10) $\equiv$ 6, 2^5 (mod 10) $\equiv$ 2, and since we have obtained another 2, we know that the pattern can do nothing else but repeat itself over and over again.

- The pattern of the units digits of the powers of 2 goes 2, 4, 8, 6, 2, 4, 8, 6, 2, 4, 8, 6, ... Each cycle of the pattern is 4 terms long.

- When two is raised to an exponent that is congruent to 1 mod 4, the units digit of the result is 2, when two is raised to an exponent that is congruent to 2 mod 4, the units digit of the result is 4, when two is raised to an exponent that is congruent to 3 mod 4, the units digit of the result is 8, and when two is raised to an exponent that is congruent to 0 mod 4, the units digit of the result is 6.

- 401 has a remainder of 1 when divided by 4, so the units digit of 2^{401} is 2.

11 Silver.

- There are 7 days in a week, and there are 365 days in a year that is not a leap year. Leap years have 366 days, and they occur every year that is a multiple of 4 (... 1996, 2000, 2004, 2008...). These are facts that should just be memorized.

- Since there are 7 days in a week, 7 days after this Wednesday will also be a Wednesday. 14 days after a Wednesday will also be a Wednesday, and so will 21 days after, 28 days after, and so forth with all multiples of 7.

- We are looking for the day of the week of 365 days after this Wednesday.

- When 365 is divided by 7, the remainder is 1. Therefore, 365 is one more than a multiple of 7. One day after Wednesday is Thursday.

12 Gold.

- We must find how many days elapse from November 29, 1972, to February 17, 1973. There are 30 days in November and 31 days in December, so from November 29 to December 31, 32 days elapse.

- There are 31 days in January, so from December 31 to January 31, 31 days elapse.

- Lastly, from January 31 to February 17, 17 days elapse. Therefore, 80 days elapse from November 29, 1972, to February 17, 1973.

- 80/7 yields a remainder of 3, and therefore 80 is three more than a multiple of 7. It follows that February 17, 1973 is three days past a Wednesday, so it is a Saturday.

- From February 17, 1973, to February 17, 1974, 365 days elapse, as there are 365 days in a year. 365 is 1 more than a multiple of 7. Therefore, February 17, 1974 is one day after a Saturday, which is a Sunday.

- In 1974, there are 28 days in February, so from February 17 to February 28, 11 days elapse. From February 28 to March 30,

30 days elapse. This is a total of 41 days that elapse between February 17 and March 30 of 1974.

- 41 is 6 more than a multiple of 7, so March 30 is 6 days past a Sunday, which is a Saturday.

13 Gold.

- Taking the value of a positive integer in mod 100 will yield the last two digits of the integer. 103^{10} mod 100 can be simplified to 3^{10} mod 100, since 103 is congruent to 3 in mod 100.

- The first four powers of 3 are 3, 9, 27, and 81. 3^5, or 81×3, has 43 as its last two digits (we do not have to worry about the hundreds digit since we are in mod 100).

- The last two digits of 3^6 are the last two digits of 43×3, which are 29. Going through with this process until we hit 3^{10}, we find that its last two digits are 49. Therefore, the tens digit of 3^{10} is 4.

14 Silver.

- What is 33 mod 15 in simplest form? It is 3, as the remainder when 33 is divided by 15 is 3. Therefore, x is not congruent to 33 in mod 15.

- What is 1589 mod 15 in simplest form? Dividing 1589 by 15 yields a remainder of 14, so 1589 is congruent to 14 in mod 15. Therefore, x is not congruent to 1589 in mod 15.

- How do we tell whether or not x is congruent to 2 mod 5? We know that x is 7 greater than a multiple of 15. We are trying to determine whether or not x is two greater than a multiple of 5.

- Every multiple of 15 is also a multiple of 5. 5 past a multiple of 15 is also a multiple of 5. This multiple of 5 is 2 less than 7 past a multiple of 15, so if x is congruent to 7 in mod 15, x is congruent to 2 in mod 5.

15 Platinum.

- Any base 8 number with digits A, B, C, D, E etc. equals $8^0 \times A + 8^1 \times B + 8^2 \times C + 8^3 \times D + 8^4 \times E. \ldots$

- All powers of 8 above 8^2 are divisible by 8^3. In mod 8^3, the number will simplify to $8^0 \times A + 8^1 \times B + 8^2 \times C$. These are its last three digits.

- $8 = 4^{3/2}$, so $8^3 = 4^{9/2}$. It follows that $n = \dfrac{9}{2}$.

Part 5: Divisibility Tricks

1 Silver.

- The square root of 161 is somewhere between 12 and 13. Therefore, we only have to test divisibility by the primes up to 12 to determine whether 161 is prime. 161 is odd, so it is not divisible by 2.

- $1 + 6 + 1 = 8$, so it is not divisible by 3. 161 does not end in a 5 or a 0, so it is not divisible by 5. However, $161/7$ does turn out to be a whole number. Therefore, 161 is composite.

2 Silver.

- What is an approximation of the square root of 1001? 30^2 is easily identified as 900, but we still have to go a little bit further.

- $31^2 = 961$, and $32^2 = 1024$, so the square root of 1001 is somewhere in between 31 and 32. To determine whether 1001 is prime or composite, we must test divisibility by all prime numbers up to 31.

- 1001 is not divisible by 2, 3, or 5, but $1001/7$ yields a result of 143. It follows that 1001 is composite.

3 Gold.

- What is an approximation of the square root of 2011? $44^2 = 1936$, and $45^2 = 2025$, so the square root of 2011 is somewhere between 44 and 45. Therefore, we must test divisibility by all prime numbers less than 44 to determine whether 2011 is prime.

- This leaves us with 14 tests to complete. In the end, we find that 2011 is prime.

4 Bronze.

- For a number to be divisible by 4, its last two digits have to be divisible by 4. 09 is not divisible by 4, so 1009 is not divisible by 4.

5 Bronze.

- For a number to be divisible by 5, it has to end in 5 or in a 0. It follows that 11,301 is not divisible by 5.

6 Silver.

- For a number to be divisible by 6, it must be divisible by both 2 and 3.

- 97, 884 is even, so it is divisible by 2.

- For a number to be divisible by 3, the sum of its digits has to be divisible by 3. $9 + 7 + 8 + 8 + 4 = 36$, and 36 is divisible by 3, so 97,884 is divisible by 3.

- Therefore, 97,884 is divisible by 6.

7 Silver.

- We must split 99 into two or more relatively prime factors and test the given number for divisibility by each of them. The best way to do this is to split 99 into 9×11.

- If this number is divisible by both 9 and 11, then it is divisible by 99. For a number to be divisible by 9, the sum of its digits has to be divisible by 9. The sum of the digits of the number is 80, so it is not divisible by 99.

8 Silver.

- Let us create a divisibility rule for 15. Splitting 15 into 3×5, we find that for a number to be divisible by 15, it has to be divisible by 3 and 5.

- This number ends in 0, so it must be divisible by 5. The sum of this number's digits is 43. 43 is not divisible by 3, so this number is not divisible by 15.

Index

Printed in the United States
by Baker & Taylor Publisher Services